Rights and Liberties in the Biotech Age

Rights and Liberties in the Biotech Age

Why We Need a Genetic Bill of Rights

Edited by Sheldon Krimsky
and Peter Shorett

A project of the Council for
Responsible Genetics

ROWMAN & LITTLEFIELD PUBLISHERS, INC.
Lanham • Boulder • New York • Toronto • Oxford

ROWMAN & LITTLEFIELD PUBLISHERS, INC.

Published in the United States of America
by Rowman & Littlefield Publishers, Inc.
A wholly owned subsidiary of The Rowman & Littlefield Publishing Group, Inc.
4501 Forbes Boulevard, Suite 200, Lanham, Maryland 20706
www.rowmanlittlefield.com

PO Box 317
Oxford
OX2 9RU, UK

British Library Cataloguing in Publication Information Available

Library of Congress Cataloging-in-Publication Data

Rights and liberties in the biotech age : why we need a genetic bill of rights / edited by
Sheldon Krimsky and Peter Shorett.
        p. cm.
Includes bibliographical references and index.
ISBN 0-7425-4340-4 (cloth : alk. paper) — ISBN 0-7425-4341-2 (pbk. : alk. paper)
1. Biotechnology—Social aspects. 2. Bioethics. [DNLM: 1. Bioethical Issues. 2. Genetic
Engineering—ethics. 3. Biotechnology—ethics. 4. Human Rights. WB 60 R571 2005]
I. Krimsky, Sheldon. II. Shorett, Peter, 1961–
TP248.23.R545 2005
174'.957—dc22   2004019684

Printed in the United States of America

⊚™ The paper used in this publication meets the minimum requirements of American
National Standard for Information Sciences—Permanence of Paper for Printed Library
Materials, ANSI/NISO Z39.48-1992.

Contents

Foreword

$\mathscr{I}$n early August 1999, a man named Max More took to the podium at the fourth annual convention of the Extropian movement, a group that includes among its advisers and luminaries many of the researchers and pamphleteers at the vanguard of the new biological (and nanotechnological and robotic) revolutions. More, who was born Max O'Connor but chose his new name to embody a philosophy of constant striving for progress, announced that he wished to propose "some amendments to the human constitution" because in many ways nature had done a poor job. His seven amendments included the elimination of death in favor of eternal life, an increase in human perception by engineering novel and improved senses, improved "emotional responses" and reshaped "behavioral patterns," enhanced memory, and improved intelligence. He sat down to the lusty cheers of his comrades.

I don't intend here to gauge the chances that the various projects he outlined will succeed, though it is worth noting that many of them have already been achieved, to one degree or another, with lab animals. (Certainly the spread of genetically engineered crops in the last fifteen years from bench experiments to continent-scale plantings gives one pause before dismissing such talk.) All I wish to draw from his enthusiasm is the sense that we stand at a threshold; in our generations we will decide whether to cross irrevocably into a very different world from the one we now inhabit.

The authors in this book, with their carefully reasoned calls for a genetic bill of rights, seem to me to be making a powerful conservative argument, and proposing amendments far more sensible, humane, and rational than the zealotry promoted by men like More. They are saying there is great value in human beings as we have known them, in plants and food crops as we have slowly and within clear boundaries developed them over millennia, in the relationship between human beings and the natural world. When I say "conservative," of course, I don't mean in the current political usage, where the term has come to mean someone who puts the endless growth of the economy above all other political goals, and is willing to countenance any disruption to that end. I mean in a more literal sense—these authors are interested in conserving much of what we intuit as the normal operation of our civilization.

That such an appeal even needs to be made underscores the dramatic nature of our moment on earth. The revelation that all life is in many ways the product of its genes is still relatively new; the notion that those genes are plastic and manipulable still seems like science fiction to most people, even as researchers prove each day that it is not. Because of that public perception, and because of the commercial pressure from people who will profit from such technologies, legislators have been slow to take up many of the questions. But that is beginning to change. With the widespread opposition to genetically modified foods in Europe, activists began to find their voices, and in many parts of the world outside the United States there has been at least some progress along the lines envisioned by the authors. Canadian authorities, for instance, have banned germline genetic modification of human beings, and the European Union has taken similar steps. Some of the rights demanded in this volume will surely be on the political agenda in years ahead: the right to exculpatory DNA evidence, for instance, and the right to genetic privacy.

But some of the rest of these questions—the rights of indigenous people, the right to live free of the threat of genotoxins, to eat untainted food, and most of all to avoid the engineered offspring envisioned by the technoutopians—are such large questions that it's possible they will never be really engaged, that the inertia of technological progress will carry us past the key thresholds before we fully realize what has happened.

Hence, it seems to me that the most valuable contribution of this volume is as a debate-starter, an identifier of the issues that simply can't be ignored, a line in the sand to defend. And though the forces against the declaration of such rights are powerful—not just the passions of the most romantic scientists, but more the forces of big business eager to profit from these advances—they are not insurmountable. Indeed, most humans are deeply uncomfortable with these technologies, sensing in their gut that the very meaning of our lives is at stake if engineering on this scale is allowed to proceed.

My only caveat would be this: that in declaring rights, which we think of primarily as individual possessions, we don't lose sight of the fact that it is as much for the sake of communities, and indeed for the commons of our own species, that this fight must be waged. People are at stake, and their liberties, but an idea is at stake as well, and that is the very idea of who we are and what it means to be a human being.

Bill McKibben
July 2004

Acknowledgments

We would like to acknowledge the efforts of the following individuals, without whom this book would not have been possible: Martha Herbert facilitated lengthy discussions among Council for Responsible Genetics (CRG) board members and rendered them into text; Claire Nader, as chair of the board, guided many feisty debates at board meetings, all of which resulted in the framing and adoption of the Genetic Bill of Rights; and Sujatha Byravan, executive director of the CRG, provided staff and office support for this book. We also express our appreciation to other members of the CRG board who participated in drafting the original Genetic Bill of Rights: Colin Gracey, Paul Billings, Philip Bereano, Debra Harry, Ruth Hubbard, Jonathan King, Sheldon Krimsky, Stuart Newman, Devon Peña, and Doreen Stabinsky. Finally, we are grateful to Eileen Smith for doing such a superb job in copyediting the manuscript.

Introduction

Sheldon Krimsky and Peter Shorett

During the past quarter century the engine of biotechnology has raced through industrial and agrarian economies like a freight train without brakes to slow it down or an engineer to steer it. The very thought of having social controls over its applications has been met by its most ardent promoters with cries of "let the free market decide." According to this view, the only justification for putting brakes on a technology would be a product that introduces a clear and present danger to human health or national security.

Biotechnology is affecting broad sectors of the economy, including agriculture, health, pharmaceuticals, the fertility industry, natural resources, material science, and forensics. For this reason, we may view biotechnology, which includes genetic and cellular engineering, as a major technological revolution. Historically, such revolutions have brought changes to society, sometimes for the better and sometimes for the worse. But we have never had a technological or political revolution that has not been accompanied by some fundamental adjustments to and controls of the forces of change. The major difference in our varied responses to technological or political change is whether we adapt to them by default or whether we make a conscious effort to take control over the possibilities.

Let us imagine a not-too-distant future when there are no state or international controls over developments in biotechnology. What outcomes might we expect? What violations of civil liberty or minority rights would be likely? Consider the following possible cases:

A witness to a crime identifies a young indigent male out of a lineup as the person who allegedly raped and murdered a young woman. A jury sentences the prime suspect to death by lethal injection. After six years of appeals, the courts uphold the sentence and the convicted felon is put to death. Subsequent to his execution, a legal team discovers that a DNA test could have proved that the wrong person was tried and convicted for the crime. But such a test was not made available to the convicted felon for lack of funds in the state budget.

In another case, the father of a young woman is diagnosed with Huntington's disease, an incurable and untreatable degenerative neurological disorder. The young woman has a 50 percent probability of contracting the disease later in life. She decides not to take the test to determine if she is carrying the Huntington's gene. Based on the uncertainty of her medical condition, her health insurer doubles her premium while her life insurance company places her in a high risk pool, which makes her insurance costs prohibitively high.

In a third case, the Organic Food Growers Association can no longer assure consumers that organic produce will be free from contamination from genetically modified strains. The reason for this is that the buffer zones that were supposed to be set up between genetically modified crops and organic crops have not been effective in preventing the contamination of organic farms. Consumers concerned about the possible spread of food allergens from mixing genes across animal and plant species or about changes in nutrient value of genetically modified crops can no longer depend on the organic label for protection.

These entirely plausible cases call out for protections against the misuse or neglectful use of new genetic technologies. This book is based on the idea that societies need a genetic bill of rights to protect the civil liberties, communitarian values, and minority rights of people against a technology that is out of control. Biotech companies are introducing untested and unlabeled genetically modified food into the marketplace. Young couples planning to have children are barraged by a multitude of new genetic tests, the consequences and value of which are not fully understood. Thousands of synthetic chemicals are exposing humans to genetic and chromosomal damage, contributing to cancer and developmental disorders. Government and state agencies are finding ways to gather DNA identification data from larger segments of the population with little concern for personal privacy. Indigenous people throughout the world are fighting multinational corporations' attempts to patent the germ plasm of their native plants and animals and their own DNA. This volume seeks to raise awareness about new genetic technologies and products that companies are introducing into the marketplace despite a widespread lack of understanding of their implications.

At the turn of the new millennium, twelve members of the board of directors of the Council for Responsible Genetics (CRG), a twenty-one-year-old public interest organization dedicated to exploring the societal implications of the new genetic technologies, voted to approve ten principles that they called the Genetic Bill of Rights. The CRG's purpose in framing these ten principles was to emphasize the link between unchecked genetic technologies and the erosion of human rights. Only a set of formal principles could properly challenge the forces of change.

There are several reasons for adopting a rights agenda to protect people from abuses resulting from genetic technologies. The first consideration takes us back to the American Bill of Rights, which was added in a series of amendments to the U.S. Constitution. When the British monarchy's rule was overthrown by the American colonists, the band of revolutionary dissidents we call "our founding fathers" formed a representative government and created the Bill of Rights to protect citizens against the tyranny of the majority. While framers of the Constitution worked out the checks and balances within the structure of government, until they adopted the Bill of Rights there was no brake against an overzealous state prepared to sacrifice individual liberties.

Of course, each of the principles in the American Bill of Rights is limited. For example, the Fourth Amendment, which secures individuals the right to be free of unwarranted government search and seizure, is not absolute. The courts have upheld involuntary searches of individuals by police and federal authorities when a suspect is believed to have committed a crime or is believed to be preparing to commit a crime. The Bill of Rights does not explicitly protect other intrusions into private affairs although the courts have found that a combination of the First, Fourth, and Fifth Amendments can be interpreted to protect certain zones of privacy.

But the adoption of a "right" establishes a burden of proof. Those who wish to violate the right must demonstrate a compelling governmental interest, such as probable cause in the case of an intrusive search or the disclosure of personal medical information, or a clear and present danger in the case of restrictions on free speech. This is why a rights approach to protecting liberties is so vital. It stands in stark contrast to debates over legislation in which those who advance a new law or regulation merely are required to justify its passage against the burdens imposed by its implementation, including its interference with free markets. In such policy decisions, fundamental rights take a backseat to the logic of cost-benefit analysis. Too often, losses that defy easy measurement are not even considered.

The American Bill of Rights is designed to protect individuals—their property, their speech, their right to worship, their right to self-protection, and their right to a fair and speedy trial. It makes no reference to the rights of common ownership, protection of public resources, or the collective public good. A different model of rights doctrines, with which the Genetic Bill of Rights (GBR) shares some characteristics, can be found in the Universal Declaration of Human Rights passed by the United Nations in 1948. Some of these rights are individuated, such as Article 5 ("No one shall be subjected to torture or to cruel, inhumane or degrading treatment or punishment") or Article 13 ("Everyone has the right to freedom of movement and residence within the boundary of each state").

But in addition to the rights of individuals, there are other rights that depend on a collective public purpose. For example, consider Article 15 ("Everyone has a right to a nationality") and Article 25 ("Everyone has a right to a standard of living adequate for the health and well-being of himself and his family"). These rights are not about preventing the state from inhibiting people to speak, act, or hold property, nor are they about interfering in one's privacy, or what we usually understand as protections against encroachment on individual freedom. Another set of rights illustrated by the UN Declaration imposes on the state an obligation to ensure the well-being of citizens, such as Article 26 ("Everyone has the right to education"). These rights are part of the commonweal function of government. The right to an education or an adequate standard of living, including health care, demands a proactive state role.

When nuclear power was unleashed, international organizations sponsored treaties on the safe uses of atomic energy. Similarly, after scientists recognized that there was a connection between chlorofluorocarbons and upper stratospheric ozone depletion, social movements helped to create an international agreement banning ozone-destroying chemicals. These treaties established a transnational agenda to protect current and future generations from the misuse of particular technologies and their despoliation of the global commons. Whether because of political, economic, scientific, or technological change, progressive communities have on numerous occasions organized themselves to counteract the threat to communitarian interests.

We can refer to these as collective rights because they depend on the role of governments to provide the social, economic, or environmental conditions that enable people to achieve a decent quality of life. Collective rights go beyond the role of the state in preventing a transgression of people's basic liberties of free speech, association, and trial by jury. Thus, the Americans with Disabilities Act gave disabled Americans new entitlements that were not afforded to them in the Bill of Rights, such as wheelchair access to public facilities. Although all rights depend on government action for their implementation, the protection of collective rights often involves an emphasis on public goods, greater allocation of state resources, and international cooperation.

In its statement of universal principles, the Genetic Bill of Rights consists both of individual and collective rights. The first article in the GBR affirms that people have a right to the preservation of the earth's biological and genetic diversity. This is a transnational right that each person can claim and a duty that all governments collectively share. Articles 2, 3, 4, 5, and 10, covering patented entities, genetically modified food, the management of biological resources, the protection from toxins, and the prohibition against prenatal genetic manipulation, are collective rights in that they refer to collective goods (the global genetic commons, the integrity of the food supply, a clean environment free of toxic substances), the preservation of which requires coordination among individuals, states, and the market. These rights also speak to a cultural ethos or an international principle of justice that has less to do with personal liberties and the inalienable rights of individuals and more to do with communitarian values. The sixth through ninth articles of the GBR, covering eugenics, genetic privacy, genetic discrimination, and DNA tests, follow in the civil liberties tradition in the American Bill of Rights.

The framework for the Genetic Bill of Rights grew out of discussions about the place of technological change in modern society and the implicit assumption among many people that there is some inexorable power vested in science that obligates societies to adapt to its discoveries. The CRG board believes this view to be patently false. Rather, it is the fundamental right and obligation of members of a democratic society to decide collectively whether a new technology is consistent with the needs and values of an informed body politic. Technology is not autonomous, but it may end up being accepted by default if society hears only the voices of its false prophets.

In cases where applications of genetic technology have resulted in public opposition, the burden is too often placed on skeptics to demonstrate why certain uses should be restricted or prohibited. The reason for this is simple. As a society we have too easily fallen prey to the idea that there is an inalienable right for science and technology to move forward. Those who question that "right" face a high burden to prove probable risks or societal costs, weighed against the economic benefits of new technologies to investors or the general public.

But not all issues raised by discoveries in genetics can be resolved by weighing traditional costs and benefits. Some of the consequences of the new genetic technologies are so significant in their social implications that they should be assessed in terms of whether they violate a basic right of individuals, minority communities, or global citizens to protect their heritage for generations to come.

The initial publication of the Genetic Bill of Rights in April 2000 included a preamble. Both the preamble and the ten principles making up the GBR are reproduced in the appendix of this volume. The original framers of the GBR understood that each principle required further analysis, justification, and interpretation. They also recognized that there would be differences in how people, even generally like-minded people, would approach the principles subsumed under the GBR.

For this volume, we invited the comments of a group consisting of CRG board members involved in the drafting of the Genetic Bill of Rights as well as others who have offered critical commentaries or participated as social activists on biotechnology issues. Their commentaries provide a historical and philosophical grounding for the GBR, but they also raise many questions.

Consider, for example, the right expressed in Article 3: "All people have a right to a food supply that has not been genetically engineered." Does this imply that there should be no genetically modified (GM) foods? Or does it imply that there should be a choice, enabling people to opt out of the global experiment on the genetic modification of the food supply if they so choose?

What responsibilities are implied in the principle that all people have a right to protection from environmental toxicants that can harm their genes or those of their offspring? After all, ultraviolet radiation from the sun is mutagenic to human DNA. How can such a right distinguish between natural and produced mutagens? With whom does responsibility rest for protecting people against environmental mutagens? And should environmental mutagens have a preferred status relative to other environmental toxicants that may harm somatic rather than germ cells (eggs, sperm, or their precursors) or that damage cells and organs without disturbing the individual's genetic makeup?

Article 10, "All people have the right to have been conceived, gestated, and born without genetic manipulation," presents its own special problems of interpretation. First, this is framed as a right to be protected against something, as opposed to a right available to someone, as in the right to take a certain action or engage in certain behavior, or hold certain beliefs. This type of right is not without precedent. Under the Occupational Safety and Health Act, workers have the right to safe working conditions. Thus, they have the right to be protected from occupational harm. A regulatory body, the Occupational Safety and Health Administration (OSHA), is supposed to protect that right for workers.

Second, how can a person have a right to be protected from something that happened to them before they became a person? This raises the question of whether rights have a retroactive function that reaches back prior to personhood. Are there such things as fetal rights? Can fetal protections be interpreted retrospectively once the fetus becomes a person?

While the GBR is framed in terms of the rights of persons, the first article suggests that people provide stewardship for all biological species on the planet. Implicitly, the interests of all species are found in the protection of biodiversity, although the concept of rights in the GBR is afforded solely to persons.

Civil society needs powerful tools to respond to the political and economic interests driving the genetics revolution. The Genetic Bill of Rights recognizes a fundamental inadequacy in mainstream utilitarian, incremental, and ad hoc approaches to protecting human rights and the environment against these formidable forces. The choices raised by genomic medicine, the transformation of our fields and farmland, and the private appropriation of our biological heritage call instead for a broad, robust decision-making framework that puts our fundamental rights and values first. Because the questions we are presented with are complex, and the prospects and perils uncertain, such a framework will undoubtedly mean both measured acceptance of some new technologies and their products, as well as outright rejection of others. In each case, however, the public deserves to know to what extent the choices our governments make in biotechnology are consistent with our fundamental liberties and the core values of our local and global communities.

Part I

BIODIVERSITY

"All people have the right to preservation of the earth's biological and genetic diversity."

Article 1 of the Genetic Bill of Rights

Genetics, "Natural Rights," and the Preservation of Biodiversity

Brian Tokar

In today's rather synthetic and sanitized human environments, it is easy to forget how dependent we are on the integrity and diversity of living ecosystems. The "web of life" appears as a mere abstraction, a paean to an earlier, perhaps simpler epoch of human history. It is only when that web is seriously damaged that we become aware of how much we rely on the rest of life on earth. Even in times of drought, floods, and wildfires, or protracted periods of unusually extreme weather, we look to our civilization's seemingly invincible technological infrastructure to protect us from nature's apparent excesses.

But our reliance on non-human natural processes persists. From the oxygen we breathe to the soil that nurtures our food, we owe our health as a species to an interconnected web of other living organisms and their almost incomprehensibly complex interrelationships. In recent decades, ecologists have demonstrated how the loss of one key species can lead to the collapse of an entire ecosystem, or how a newly introduced species can become so invasive that everything around it is altered by its presence.

For agrarian peoples and the surviving indigenous communities all over the earth, this reliance on natural processes is far more immediate. Not long ago, scientists discovered that indigenous corn crops in southern Mexico—the biological center of origin for domesticated corn—were contaminated with pollen from imported genetically engineered corn varieties.[1] This was one event that highlighted the fragility of long-established patterns of living on the land, in the face of genetic technologies that by their very design aim to flout natural limits.

Is there a fundamental right to enjoy the fruits of an intact biodiversity? Who defines and establishes such a right? Does it apply equally to everyone?

11

Do other living beings have rights comparable—or perhaps even superior—to those of humans? Such questions have intrigued philosophers and theologians for centuries. With the rise of modern environmentalism—and a technological infrastructure capable of previously unimaginable acts of destruction—they have become increasingly urgent.

In the Anglo-American tradition, discussions of "natural rights" are usually traced back to the works of John Locke. But to Locke, nature was of interest mainly as property. The defense of private property extended even to owners of slaves, and legal rights were a privilege limited to an educated elite. It took several generations of English and American radicals to transform Locke's rights discourse into a more universalist embrace of the fundamental "rights of man." Environmental philosophers such as Roderick Nash have celebrated the gradual opening of the discourse of natural rights to include Native Americans, African slaves, wild as well as domesticated animals, and, ultimately, all of life on earth.[2] From biblical entreaties about respect for strangers and the proper care of God's creation, to seventeenth-century debates over vivisection and the treatment of domestic animals, to Aldo Leopold's twentieth-century "land ethic,"[3] discussions about the ethics of humanity's relationship with non-human nature have been fraught with ambiguity as well as hope. What are "natural rights" for, and what exactly, do they protect?

In today's legal system, burdened as it is by two hundred years of capitalist property relations, discussions about the preservation of nature are too often transformed into debates about its proper use, ownership, and sale. At the turn of the twentieth century, John Muir's visionary appeals to the integrity of creation gave way to a "progressive" utilitarianism that sought to rationally manage the "taming" of the American West.[4] A century later, efforts to enshrine the protection of biodiversity in a United Nations treaty led to the adoption of a Convention on Biological Diversity that instead established the rights of the biotechnology industry to manage the ownership of biodiversity.[5]

The 1980s and 1990s were filled with efforts to link the preservation of biodiversity to private ownership of land and resources. Environmental groups such as Conservation International and the World Wildlife Fund supported so-called "debt-for-nature swaps," through which countries in Latin America and Africa would be allowed to write off a small portion of their mushrooming national debt in exchange for commitments to "sustainably manage" ecologically sensitive areas. Superficially, it appeared that both sides—and the cause of natural preservation—would benefit from this arrangement but, in practice, a series of rather predictable problems persisted. Often, buffer zones surrounding the protected areas were subjected to intensified extraction of timber and other resources, and indigenous peoples living in sensitive zones were either forcibly evicted or compelled to subdivide communal landholdings into individual private plots.

In 1991, the pharmaceutical giant Merck signed an agreement with Costa Rica's National Biodiversity Institute (INBio) offering the institute $1 million over two years for the exclusive right to screen native species for possible new drugs.[6] The arrangement was touted as a model of sustainable development, with INBio promised a share of royalties on any new products that were discovered there. Five years later, researchers found that continuing support from Merck was being used almost exclusively to support bioprospecting activities in Costa Rica's conservation areas and some royalties were going toward routine maintenance of national parks; none was going to conservation.

In 2002, Conservation International (CI) collaborated with the U.S. Agency for International Development (USAID) to identify thirty-two communities of Mayan people that had settled on the edges of the Montes Azules Biosphere Reserve, a vast expanse of lowland rainforest in Mexico's southern state of Chiapas. USAID provided sophisticated aerial photography equipment to CI that could detect individual inhabitants and their simple household goods. The settlements had mostly been established during the 1970s, when the Mexican government encouraged refugees from the region's huge, quasi-feudal coffee plantations to settle in the rainforest to relieve land pressures in the Chiapas highlands. But now, according to CI and the Mexican government, they represent a threat to the rainforest and have to be forcibly removed.

As the area has become increasingly militarized in recent years, with the construction of two huge military bases, Mexican officials are seeking to wrap themselves in the mantle of environmentalism. Former president Ernesto Zedillo declared the Montes Azules reserve a "global public good" to be conserved for all of humanity. "The destruction of forests is a worse crime than terrorism or narcotrafficking," says Mexico's new UN ambassador. One community has already been forcibly expelled from the rainforest, and many others have been threatened by armed government agents.

But who can best be relied on to take care of the land and protect its biodiversity? The Mexican government has embarked on a mega-project, termed the Plan Puebla Panama, which seeks to transform the entire region, from southern Mexico all the way to Panama, into a single Free Trade Zone, complete with high-speed rail corridors, container ports, and vast export processing complexes. Conservation International has established several biological research stations in the Montes Azules in collaboration with the Mexican-based transnational Grupo Pulsar, a company with interests in large-scale timber extraction and the sale of genetically engineered seeds.

Meanwhile, an environmental delegation to the Montes Azules in the spring of 2003 discovered that several communities have undertaken efforts to demonstrate highly sustainable ways of living on the land, including an end to slash-and-burn agriculture and the use of agricultural chemicals.[7] "These forests are for the monkeys, birds, and jaguars," explained the caretaker of a lush cornfield hidden underneath the forest canopy near the settlement of Nuevo San Gregorio. "They are also important to our future. If we chop down the trees, our children won't know what the mountains look like."

"The experts in Mexico City know about the thinning of the ozone layer, but they can't even see the stars at night," explained another leader of the community. "They can't see that it's the people with the Mayan faces who are taking care of the land."

Stories like these add layers of uncertainty to simple notions of biodiversity as the "common heritage" of humankind. With vast discrepancies of power and wealth throughout the world today, it is becoming increasingly difficult to realize the ideal of a universal human interest. Differences in wealth and life experience breed vastly different relationships to the land, personal values, and ethical systems. As the gap widens, it becomes increasingly difficult for privileged people in the industrialized world to understand what the world looks like to those whose survival depends intimately on the integrity of the forests, grasslands, and mountainsides where they actually live.

The commercial production of over one hundred and fifty million acres of genetically engineered crops represents one of the severest tests to date of our ability to conserve biodiversity. Genetic engineering is an assault on the integrity of the proverbial web of life. By forcibly injecting new combinations of genetic information into the chromosomes of plants and animals, and then releasing these new organisms into the environment, genetic engineers are risking unprecedented genetic and ecological disruptions. Genetic engineering is an inherently uncertain, and therefore inherently unpredictable, process, interfering with the ability of living cells to properly regulate cellular processes. We see a host of unexpected consequences, such as the silencing of genes, the production of novel proteins, and the enhancement or suppression of physiological functions that may have no discernible relationship to the particular trait that scientists are attempting to alter.[8]

Some proponents of genetic manipulation argue that GMOs are increasing biodiversity by adding new, artificial combinations of genes to the gene pool. This spurious claim speaks volumes about reductionist science's growing distance from a nuanced appreciation and understanding of natural processes. The genetic makeup of today's living organisms is the product of billions of years of biological evolution. It is the result of a process that is self-organizing and internally regulated to a truly astonishing degree. Living cells have evolved means of DNA repair that guard against genetic instability, and genetic engineers employ a host of genetic "tricks"—using promoter sequences from viruses, for example—in their attempt to override these controls. The resulting "novel organisms" are often inherently unstable and have properties that

cannot be reliably predicted or controlled. A science that would seek to protect, and sometimes enhance, biodiversity would be far less interventionist, merging systematic biological knowledge with a far more evolved sensitivity to the inner logic of living processes.

It is the task of those of us in the global North who are committed to the preservation of biodiversity to stretch the limits of our ethical understanding and our concept of scientific knowledge. This would include a far greater appreciation for the worldviews of indigenous communities, whose values have evolved in a far more participatory relationship with the living world. The aim is not to romanticize the lives of these people, nor their diverse understandings of the world. Rather, they help us reach beyond the attitude of managerial detachment that has made environmentalists unwittingly complicit in the abuse of people and their land, and driven genetic engineers to replace natural diversity with a simplified world of artifice and manipulation. We must forthrightly reject the commodification of the earth, of plants and animals, and of genetic information. We can celebrate human creativity as well as our interdependence, while challenging the economic and political institutions that continually enforce norms of neglect, abuse, and exploitation. We can begin a process of ethical and social transformation, rooted in new institutions of cooperation, mutual aid, and complementarity that help realize the dream of human freedom and make the preservation of the earth's biodiversity a daily, very tangible reality.

NOTES

1. David Quist and Ignacio H. Chapela, "Transgenic DNA Introgressed into Traditional Maize Landraces in Oaxaca, Mexico," *Nature* 414 (November 29, 2001): 541–3.

2. Roderick F. Nash, *The Rights of Nature: A History of Environmental Ethics* (Madison: University of Wisconsin Press, 1989).

3. Aldo Leopold, "The Land Ethic," in *A Sand County Almanac* (New York: Ballantine Books, 1970), 237–64 (original edition, London: Oxford University Press, 1949).

4. See for example Edwin W. Teale, *The Wilderness World of John Muir* (Boston: Houghton Mifflin, 1954); on the conflict between romanticism and utilitarianism in the American West, see Donald Worster, *Nature's Economy* (San Francisco: Sierra Club Books, 1977).

5. Brian Tokar, "After the 'Earth Summit,'" *Z Magazine* (September 1992): 8–14; Brian Tokar, "Environmentalism, Clinton Style," *Z Magazine* (October 1993): 30–35. The Convention's official website is www.biodiv.org.

6. An overview of the Merck/INBio agreement is available at www.wri.org/wri/biodiv/b34-gbs.html.

7. The delegation, sponsored by Global Exchange, issued a final declaration on March 14, 2003; the text is available at www.wrm.org.uy/bulletin/68/Mexico.html.

8. These consequences are summarized in Jeffrey M. Smith, *Seeds of Deception* (Fairfield, IA: Yes Books, 2003): 47–75. For fuller descriptions, see Evelyn Fox Keller, *The Century of the Gene* (Cambridge, MA: Harvard University Press, 2000), and Mae-wan Ho, *Genetic Engineering: Dreams or Nightmares?* rev. ed. (London: Continuum Publishing Group, 2000). For the cautionary tale of the first genetically engineered pig, see Andrew Kimbrell, *The Human Body Shop: The Engineering and Marketing of Life* (San Francisco: Harper Collins, 1993), 175–6.

• *2* •

The Right to Biodiversity: A Concept Rooted in International Law and Understanding

Philip Bereano

At the Rio Earth Summit in June 1992, the nations of the world adopted Agenda 21, which declared:

> Our planet's essential goods and services depend on the variety and variability of genes, species, and populations and ecosystems. Biological resources feed and clothe us and provide housing, medicines and spiritual nourishment. The natural ecosystems of forests, savannahs, pastures and rangelands, deserts, tundras, rivers, lakes and seas contain most of the Earth's biodiversity. Farmers' fields and gardens are also of great importance as repositories, while gene banks, botanical gardens, zoos and other germplasm repositories make a small but significant contribution. The current decline in biodiversity is largely the result of human activity and represents a serious threat to human development.[1]

Then most of the nations of the world signed on to a Convention on Biological Diversity (CBD), writing in the preamble that they were:

> Conscious of the intrinsic value of biological diversity and of the ecological, genetic, social, economic, scientific, educational, cultural, recreational and aesthetic values of biological diversity and its components,
> Conscious also of the importance of biological diversity for evolution and for maintaining life sustaining systems of the biosphere, . . .[2]

In the past quarter century, "biodiversity" has evolved from a somewhat arcane term to a central concept in twenty-first-century environmentalism. Perhaps this is because it is so expressive of ecological holism. Or its importance may be due to the precariousness of continued diversity in the natural

17

world. Among the threats to biodiversity are the over-exploitation of land and seas for short-term financial gain (e.g., depletion of fish stocks), corporate globalization (facilitating assaults on nature on a scale previously unimaginable), global climate changes resulting from human activities, and the trangenic engineering of novel organisms. This last aspect serves as the premise of the Cartagena Biosafety Protocol, namely "the rapid expansion of modern biotechnology and the growing public concern over its potential adverse effects on biological diversity,"[3] which became effective in September 2003.

This linkage between genetic engineering and natural diversity is the underlying rationale for the first of the CRG's rights in this Genetic Bill of Rights. The overall assault on biodiversity is based on elements of reductionism (genomics as a LEGO toy) and dominance (of humans over nature, capitalism over indigenous collectivism, and the industrial paradigm of the North over all cultural works of the South). As our colleague Vandana Shiva has taught us, the reductionist biology behind modern genetic engineering "declare(s) organisms and their functions useless on the basis of ignorance of their structure and function."[4] As Shiva explains this approach, all nature is inert, ironically dead, and literally useless unless it is commercially exploitable.

Thus, the "diverse knowledge systems (of women, non-Western peoples, and local ecosystems) were not treated as legitimate ways of knowing."[5] Only commodification signifies value. So growers of some of the 17,000 distinct varieties of rice were pressured by the Indian government, USAID, and the Rockefeller Foundation to switch to Green Revolution rice that could enter world commodity distribution. And the recalcitrant seed bank bureaucrat who refused to be a cheerleader for the disappearance of these landraces was fired as an obstacle to progress. Meanwhile, fields containing scores of plant varieties, used for millennia for food, fodder, and fiber, have been plowed under as "weeds" to make way for the new "miracle" varieties. Eliminating biodiversity has been quite profitable for agribusinesses, selling the required chemical inputs as well.

The resulting monocultures of the field correspond to what Shiva has called "monocultures of the mind," a totalitarian ideology that privileges uniformity over unity, identical replicants over the diversity of a natural system operating in ecological balance. Predictability is prized; surprise and wonder are banished.

> Homogenization and monocultures introduce violence at many levels. Monocultures are always associated with political violence—the use of coercion, control and centralization. Without centralized control and coercive force, this world filled with the richness of diversity cannot be transformed

into homogeneous structures, and the monocultures cannot be maintained. Self-organized and decentralized communities and ecosystems give rise to diversity. Globalization gives rise to coercively controlled monocultures. Monocultures are also associated with ecological violence—a declaration of war against nature's diverse species.[6]

It is somewhat astonishing, in this context, that 180 countries (not including the United States) have become Parties to the UN Convention on Biological Diversity.

> The objectives of this Convention, to be pursued in accordance with its relevant provisions, are the conservation of biological diversity, the sustainable use of its components, and the fair and equitable sharing of the benefits arising out of the utilization of genetic resources, including by appropriate access to genetic resources and by appropriate transfer of relevant technologies, taking into account all rights over those resources and to technologies, and by appropriate funding.[7]

Thus, the threefold goals of the CBD are conservation, sustainable use, and the sharing of the benefits of biodiversity by all nations and communities, although in reality these goals are sometimes in conflict and sometimes overwhelmed by outright commercial exploitation. However, over one hundred countries have developed nature biodiversity strategies and action plans; how strong these are and how faithfully they are being implemented are worthwhile subjects for further research.

Under the Convention, a periodic progress report, *Global Biodiversity Outlook*, is published which is recommended to include "an analysis of the steps being taken by the global community to ensure that biodiversity is conserved and used sustainably, and that the benefits arising from the use of genetic resources are shared equitably."[8]

Suggestions of how biodiversity might be protected, such as by "the participation and support of local communities" were provided at Rio, both in Agenda 21 and the CBD itself.[9] In addition, it was emphasized that "States have the sovereign right to exploit their own biological resources pursuant to their environmental policies, as well as the responsibility to conserve their biodiversity and use their biological resources sustainably, and to ensure that activities within their jurisdiction or control do not cause damage to the biological diversity of other States or of areas beyond the limits of national jurisdiction."[10]

The linkage to a genetic bill of rights is provided in the same document where it is noted that "recent advances in biotechnology have pointed up the likely potential for agriculture, health and welfare and for the environmental purposes of the genetic material contained in plants, animals, and micro-organisms." Thus, Article 10 of the Convention obligates the Parties to:

(a) Integrate consideration of the conservation and sustainable use of biological resources into national decision-making;
(b) Adopt measures relating to the use of biological resources to avoid or minimize adverse impacts on biological diversity;
(c) Protect and encourage customary use of biological resources in accordance with traditional cultural practices that are compatible with conservation or sustainable use requirements;
(d) Support local populations to develop and implement remedial action in degraded areas where biological diversity has been reduced; and
(e) Encourage cooperation between governmental authorities and the private sector in developing methods for sustainable use of biological resources.

However, in summing up activities under the CBD after the Seventh Conference of the Parties in February 2004, with a continuing loss of biodiversity estimated at 60,000 species per year, civil society organizations have been largely critical. Claims that "the World Trade Organization [is] finally in charge of the Convention" and demands that "all governments must act with urgency to reduce, even eliminate, the threats to biodiversity loss and implement the precautionary principle" were common.[11] After ten years, it is clear that the CBD is not a magic bullet for the conservation of biological diversity, nor does it guarantee the improvement of the rights and roles of indigenous peoples and communities.[12]

The reasons for declaring that all persons have an ongoing "right" to biological and genetic diversity may be based on a number of grounds—utilitarian economics, biological realities, and lyrical philosophy.

ECONOMIC

In the words of Agenda 21, "Biological resources constitute a capital asset with great potential for yielding sustainable benefits."[13] The economic value of biodiversity has been made increasingly explicit.[14] The discussions at the UN's World Summit on Sustainable Development in Johannesburg in 2002 frequently recognized that conservation and development can no longer be seen as conflicting goals but are instead mutually interdependent; biological diversity must now be considered an essential part of efforts to eradicate poverty and achieve sustainable development: "Human activities are having an increasing impact on the integrity of ecosystems that provide essential resources and services for human well-being and economic activities. Managing the natural resources base in a sustainable and integrated manner is essential for sustainable development."[15]

The economic basis for the CBD is a trade-off in which "access to genetic resources" (overwhelmingly in the Third World) will be exchanged for a "sharing of benefits" and technology transfer. In other words, the South had enough genetic leverage to put a formal stop to a continued rip-off, but their bio-riches are not off-limits to transnational corporations. Hopefully, some profits will flow South. However, the CBD makes the nation-state, not the indigenous communities within its bounds, the owner of the biodiversity. Thus, it is unclear who will ultimately benefit. The CBD is infused with industrial society's values; bioprospecting and patenting (with the payment of royalties) are compatible with its technophilic posture.[16]

BIOLOGICAL

The 2001 *Global Biodiversity Outlook* makes clear the biological rationale for preserving biodiversity:

> Genes provide the blueprint for the construction and functioning of organisms, and their diversity is thus clearly fundamental. The Convention puts due emphasis on genetic resources, i.e. the genetic diversity responsible for key properties of organisms used by humans, for food, medicine or other purposes, and which provides the potential for future modifications to these organisms. However, genes in nature are expressed only through the form and differential survival of organisms, and if attempts are made to manipulate genes, for example in bioengineering, it is important to focus on the requirements of whole organisms if this is to be undertaken successfully. Accordingly, the diversity of organisms tends to be central to biodiversity studies, and species diversity is a generally useful and practicable measure of biodiversity.
>
> At the same time, in seeking to make management intervention as efficient as possible, it is essential to take an holistic view of biodiversity and address the interactions that species have with each other and their non-living environment, i.e. to work from an ecological perspective.[17]

Thus, ecological diversity is increasingly appreciated for its own inherent value. This biological raison d'être continues to be a lively component of policy analysis. If we want to have biological life tomorrow, we must protect it today. Diversity signifies ecological health and serves to ensure that species survive through difficult times (e.g., climatic stress). The functioning of viable ecosystems is the underpinning of sustainable development and the mainte-nance of human life and the quality of that life.

LYRICAL

A major thrust behind the evolving philosophical bases of biodiversity is a dra-matic reversal of the exploitative relationship to nature that has been a hallmark of industrial society. Vandana Shiva noted in her book *Biopiracy* that "conser-vation of biodiversity . . . involves the conservation of cultural diversity and a plurality of knowledge traditions."[18] In 1991, an International Seminar on Sus-tainable Development, held in Forteleza, Brazil, by the Esquel Foundation and the State of Ceará, proclaimed, "Sustainable development is not a means to en-sure the continuous exploitation of natural resources, with the traditional dis-regard for equity considerations. It is an end in itself, since it implies the final objective of improving the quality of life of people, on a permanent basis."[19]

In all cultures, nature has stimulated rhapsodic awe, infusing literature, art, music, and dance with its images: fecundity, balance, purposefulness. Interna-tional documents such as Agenda 21 have, in their dry prose, also celebrated the cultural importance of biodiversity:

> Recognize and foster the traditional methods and the knowledge of in-digenous people and their communities, emphasizing the particular role of women, relevant to the conservation of biological diversity and the sustain-able use of biological resources, and ensure the opportunity for the partici-pation of those groups in the economic and commercial benefits derived from the use of such traditional methods and knowledge.[20]

> Subject to national legislation, take action to respect, record, protect and promote the wider application of the knowledge, innovations and practices of indigenous and local communities embodying traditional lifestyles for the conservation of biological diversity and the sustainable use of biological resources, with a view to the fair and equitable sharing of the benefits aris-ing, and promote mechanisms to involve those communities, including women, in the conservation and management of ecosystems.[21]

True sustainability requires thorough democratization of decision-making and the provision of adequate resources to all. In other words, a good society is the best technology. In our era of powerful corporate globalization, represented by the non-sustainable priorities of the World Trade Organization, the World Bank, and the International Monetary Fund (being heroically resisted, one must note), it is not at all clear that true biological and social sustainability will emerge.

Whether these innovative ways of thinking, presented by social activists such as Shiva, can really help in some of the conflicts between indigenous and "modern" cultures is still problematic. A colleague in the 1970s told of his work on siting a liquid natural gas port on the California coast. The promontory that was "rationally" determined to be the best location was also the site at which a local tribe of Native Americans believed the spirits of the dead departed this earth and entered eternity. How to evaluate as an impact the perpetual entrapment of dead souls on this earth (and presumably their higher concentration at the California coastline as they jockey for the now-blocked portal) is still a bit mind-boggling. But then, we would not build a supermarket on Calvary, I suppose, even if we didn't personally believe in Jesus' resurrection.

> In the era of genetic engineering and patents, life itself is becoming colonized. Ecological action in the biotechnology era involves keeping the self-organization of living systems free—free of technological manipulations that destroy the capacities of communities to search for their own solutions to human problems from the richness of the biodiversity that we have been endowed with.[22]

Our massive disturbance of the earth's diverse biosystems has produced novel allergies, diseases jumping between species, superweeds, and the irreparable loss of thousands of species. We cannot know whether any of these might have been economically valuable (a disease cure, perhaps) or critical to the healthy maintenance of an ecosystem, or whether its sheer existence and transcendental beauty would have been a continued delight to the human mind and soul.

Sinister forces are committed to reducing the world's biodiversity, for narrow and selfish reasons. In Johannesburg, Emmy Simmons, the assistant administrator of USAID, challenged those of us defending biodiversity by exalting that "in four years, we will have planted so many GMOs in South Africa that the gene pool of the entire continent will be contaminated by transgenes."[23] Such profits by Monsanto may give pleasure to its sycophants but would certainly be a death knell to "the right of all people to the preservation of the earth's biological and genetic diversity" as recognized by the CRG.

The inclusion of a right to biodiversity in this CRG document, and its positioning as the first in the listing of rights, reflects this evolution of international thinking—by authors, social activists, bureaucrats, and politicians—that ecological realities are essential for human existence. That this fundamental epistemological verity has found expression in the mundane language of international treaties and reports is both a measure of how profoundly it is affecting human thinking and how fundamental biodiversity is to our very existence.

> All leaves are this leaf,
> all petals, this flower
> in a lie of abundance.
> All fruit is the same,
> the trees are only one tree
> and one flower sustains all the earth.
>
> Pablo Neruda, "Unity"

NOTES

1. United Nations Division for Sustainable Development, *Agenda 21*, text at www.un.org/esa/sustdev/documents/agenda21/english/agenda21toc.htm (adopted June 1992, Rio de Janeiro).

2. United Nations Environmental Program, *Convention on Biological Diversity*, text at www.biodiv.org/convention/articles.asp.

3. Convention on Biological Diversity (CBD), *Cartagena Protocol on Biosafety*, text at www.biodiv.org/biosafety/protocol.asp, preamble.

4. Vandana Shiva, *Biopiracy: The Plunder of Nature and Knowledge* (Boston: South End Press, 1997), 22.

5. Shiva, *Biopiracy*, 24.

6. Shiva, *Biopiracy*, 101–2.

7. United Nations Environmental Program, *Convention on Biological Diversity*.

8. CBD, *Global Biodiversity Outlook* (November 2001), at www.biodiv.org/gbo/.

9. United Nations Environmental Program, *Convention on Biological Diversity*, n. 1, sec. 15 (3).

10. United Nations Environmental Program, *Convention on Biological Diversity*, n. 1, sec. 15 (3).

11. ECO, "The Voice of the NGO Community in the International Environmental Conventions," www.itdg.org.

12. ETC Group, *Communiqué*, no. 83 (January/February 2004), www.etcgroup.org.

13. CBD, *Global Biodiversity Outlook* (November 2001).

14. "The genes, species and ecosystems that comprise biological diversity provide resources and services that are essential to mankind. All sectors of world society affect this diversity to a greater or lesser extent, whether through direct exploitation of resources or the indirect impact of other activities. Different cultures and societies use, value and protect these resources and services in a variety of ways. Their capacity to manage and benefit from biological diversity also varies considerably, because of location, state of development and differential access to the information and technology needed."

15. World Summit on Sustainable Development, *Plan of Implementation*, IV, sec. 23, www.johannesburgsummit.org/html/documents/summit_docs/2309_planfinal.htm.

16. United Nations Environmental Program, *Convention on Biological Diversity*.
17. CBD, *Global Biodiversity Outlook*, chapter 1.
18. Shiva, *Biopiracy*, 123.
19. International Seminar on Sustainable Development (Forteleza, Brazil, 1992).
20. UN Division for Sustainable Development, *Agenda 21*, sec. 15 (4) (g).
21. UN Division for Sustainable Development, *Agenda 21*, sec. 15 (5).
22. Shiva, *Biopiracy*, 39.
23. Personal communication.

Part II

LIFE PATENTS

"All people have a right to a world in which living organisms cannot be patented, including human beings, animals, plants, microorganisms and all their parts."

Article 2 of the Genetic Bill of Rights

Life Patents and Democratic Values

Matthew Albright

$\mathscr{P}$atents on life—intellectual property rights on human beings, animals, plants, microorganisms, and all their parts—are a fairly recent phenomenon, but the history of patents goes back centuries. The patent system in the United States was originally set up to fulfill a constitutional mandate to promote science and to protect inventors and artists. Ironically, the writers of the Constitution were careful not to use the word "patent" in their document because they knew it would arouse suspicion. The fact that patents had been used by the British government to monopolize such basic staples as salt, oils, vinegar, and starch was still fresh in the colonists' minds. Earlier, Christopher Columbus used the word to describe the islands that he discovered, and to claim the land, the resources, and the actual people he saw as property of the Spanish crown.

When considering the policy of awarding patents on life, we would do well to go back to the intent of the Constitution in setting up the patent system. When one compares the constitutional mandate with the current policy of awarding patents on life, however, it is clear that there is a disjuncture between the original intent of the patent system as outlined in the Constitution and how the patent system is now being used. Further, we find that the policy of awarding life patents was arrived at without public or congressional input, or even a decision from the executive branch. In essence, the policy of awarding life patents was arrived at undemocratically and without the advantages of open debate as guaranteed by the right to free speech. Ultimately, we need go no farther than the Constitution and the Bill of Rights for guidance on developing a policy on property rights that covers human beings, animals, plants, microorganisms, and all their parts. As well, within the Constitution and the

Bill of Rights, we find pointers for creating governmental bodies—alternatives to the U.S. Patent and Trademark Office—to carry out the constitutional mandate.

This chapter will first analyze patents on life in light of Article I, Section 8, of the U.S. Constitution. Even a cursory survey will show that life patents go against two of the main intents of the patent system as outlined in the Constitution: promoting the progress of science and protecting the rights of inventors. Second, this chapter will demonstrate that the current policy of granting patents on life fails democratically when one examines the process that was used to arrive at that policy.

PROMOTING THE PROGRESS OF SCIENCE AND USEFUL ARTS

"Congress shall have the power," says Article I, Section 8, of the U.S. Constitution, "to promote the progress of science and useful arts, by securing for limited times to authors and inventors the exclusive right to their respective writings and discoveries." From this relatively short phrase the U.S. Patent Office was created.

Does the Patent Office, with its policy of granting intellectual property rights on molecular biological material, "promote the progress of science and useful arts," as the Constitution mandates? We might start with the broader question of whether a system of awarding patents as a whole promotes the progress of science and protects the property of inventors. Is the patent system the most efficient way of supporting technological advancement in a given society? Or does a patent system create monopolies on basic knowledge and materials and ultimately stifle innovation, as it once did when used by the British crown?

This was certainly a question that Thomas Jefferson and other founding fathers asked themselves when considering how to implement Article I, Section 8, of the Constitution. For obvious reasons, the American people considered patents to be highly political issues. Jefferson, the first director of the Patent Office, approached every patent application with a cautiousness toward monopolies and due consideration for the public good. He personally tested many of the inventions himself and, in his first year managing the office, granted only three patents.[1] As well, although Jefferson invented many things, he himself never applied for a patent.[2] Nor did another famous colonial inventor, Benjamin Franklin. Their reasons for not applying were simple: they believed that their technical inventiveness should be shared as ideas, allowed to be freely tested and improved upon by other inventive minds.[3]

Today, critics of the patent system point to Japan's postwar technological progress in which a lack of patent protection was key to the country's success. Proponents of the patent system use U.S. history to demonstrate that this country owes its formidable technological strength to a strong patent system. This argument can be countered by noting that U.S. technological leaps in many cases, most notably within the aerospace and automobile industries, came about only after the stringent patent laws were loosened. Indeed, the patent system in some instances was blamed not only for hindering scientific advancement but for stalling the national economy as well. The aeronautic industry in the United States would not have experienced the success it had between the world wars if the Wright brothers' original airplane patent was allowed to be enforced. Instead, as also occurred in the early years of the automobile industry, the aeronautic industry was forced by the government to share the upstream knowledge and technology that specific individuals and companies had monopolized through patents.[4] In 1938, President Franklin D. Roosevelt went so far as to tell Congress that patents were one of the causes of the "economic malaise gripping the country."[5]

Academics, industry representatives, and policy makers continue to question whether the patent system is the best vehicle for promoting the progress of science and the useful arts, that is, technology. The most recent addition to the debate comes from the Commission on Intellectual Property Rights (CIPR), a body of lawyers, economists, and scientists assigned by the British government to look at how IPRs function in developing countries. Their September 2002 report found that although the Western-style patent system works well in some cultural contexts, it clearly stifles innovation and competition in others.[6] "The central message," the *Economist* reported on the CIPR document, "is both clear and controversial: poor places should avoid committing themselves to rich-world systems of IPR protection unless such systems are beneficial to their needs."[7]

Perhaps the best that can be said is that in some contexts the patent system promotes scientific and technological advance and in others it does not. Throughout U.S. history, the patent system at times stimulated innovation but, at other times, clearly hindered it, and industry found itself in dire need of political intervention.

With regard to patents on life, the patent system clearly *does not* promote scientific progress. Extensive studies have been conducted on how intellectual property rights on biological material restrict the sharing of information between scientists. A 2002 survey found that 47 percent of university geneticists had been denied access to other scientists' research results.[8] An earlier survey found that genetic researchers restrict free exchange of research results more than all other life science researchers. Nearly 20 percent of all life science researchers surveyed reported delaying the release of their own research results by more than six months. The main reason given was the need to wait for a patent application.[9] The patent policy on molecular biological material has led to less collateral testing of research results, incomplete disclosure of results, and hiding of research findings. Progress is stifled and science itself is failing.

An oft-repeated argument is that gene patents promote scientific progress because they give an incentive for financial investment in biotechnology projects. Although this is not the place for an in-depth response to that question, the argument certainly must be reexamined in light of the recent downturn in the biotechnology industry and the stock market in general. Far from spurring innovation and economic development, life patents appear to be one of the causes of the biotechnology industry's recent financial slump.

The biotechnology industry is in trouble because it has yet to make a real profit, and it has not made a profit because, as an industry, it has produced very few marketable products. In the words of John Wilderson of the investment company Galen Associates, "A biotech company is a pharma company without sales."[10]

As with the early aerospace industry, we are amazed and excited by the new discoveries in genetic research, but we have yet to see any successful products from that research. "Unfortunately, the deluge in data [in genomics] has yet to spur any dramatic increase in the number of new drugs discovered," the *Economist* commented in 2001.[11]

For the most part, the industry has survived through announcements of its discoveries of upstream knowledge—and subsequent life patents on those discoveries—that buoyed stock prices and interested venture capitalists. Up until now, the biotechnology industry has been an industry of knowledge, producing maps of genomes and descriptions of genetic processes and structures. The industry has also developed genetic tests and genetic databases, but these knowledge-based products have very little use except in the context of further research on disease. The biotechnology industry has bought and traded this knowledge of human biological structures through the use of life patents. It has even made money by licensing and selling these patents to other researchers, and, in some cases, biotechnology companies have even profited from patent litigations. As exemplified disastrously by the dot-com industry, however, an industry cannot survive long without real products, real customers, and real profits.

If biotechnology companies are to survive, they must make the jump from an era of inventiveness to an era of industry, producing real products that yield real profits. Those products will probably be in the form of drugs. Standing in the way of those drugs, however, is a minefield of life patents that monopolize basic knowledge that needs to be shared in order for real products to be produced.

THE INVENTOR'S EXCLUSIVE RIGHT
TO INVENTIONS AND DISCOVERIES

It is questionable whether the policy of granting patents on life fulfills the second set of requirements of the constitutional mandate: that Congress secure "for limited times to authors and inventors the exclusive right to their respective writings and discoveries."

We may pause here and consider the argument that life forms and genes are not invented by anyone, no matter how genetically modified they are, but are products of evolution over thousands of years. Given that the Patent Office considers both discoveries and inventions as permissible criteria for getting a patent, we can leave this argument aside and ask, even as a discovery, whether any one scientist or company can be credited with discovering or inventing such material.

Most of the discoveries in life science are built on years of cooperative research among many different groups and scientists. Some discoveries can be traced to very specific groups that did the major legwork. Even in those instances, however, the scientists are not getting the credit for their work through patents. The patents, instead, are going to companies who may have done some work on the discovery but have been awarded the patent because of an efficient and powerful legal team.

The search for the so-called breast cancer gene is a case in point. Marie Claire-King and many other scientists from around the globe spent fifteen years searching for a genetic mutation that appeared to have a correlation to breast cancer. Building on their extensive research, Myriad Genetics made a successful mapping of the mutation in 1994 and to that company the patent was given. As Dr. Donald Bruce, director of the Society, Religion, and Technology Project, put it: "Years of patient research by several groups got close to locating the first gene [for breast cancer], but at the last minute a newly set-up U.S. company hustled in, did the last few steps and claimed the whole gold seam as their own private property."[12]

In other cases, private companies or universities have patented genetic material about which they know very little. In an infamous case, Human Genome Sciences (HGS) was granted a patent on a cell surface receptor called CCR5 just months after scientists at the National Institutes of Health discovered a relationship between CCR5 and HIV/AIDS.[13] Though HGS was unaware of CCR5's connection with HIV/AIDS, the company's stock shot up after the patent was granted, and HGS continues to own the property rights to CCR5. In the cynical words of one biotech lawyer, "scientific credit is one thing, patent law is something else."[14]

Other patents on life, such as those that cover naturally occurring plants and foodstuffs (e.g., rice, quinoa, and Indian neem), are little more than blatant piracy from cultures whose history has long demonstrated the "utility" of such plants. The use of these plants over the course of thousands of years refutes the "inventiveness" claimed by universities and companies in their patents.

U.S. patent policy under the auspices of the U.S. Patent Office can only be viewed as unconstitutional when we look at the rather simple intents of the Constitution in forming it: promoting the progress of science and protecting inventors. The Patent Office does not consider the broader context in which an "invention" is proposed, including whether intellectual property on such inventions actually hinders scientific progress or whether the scientist or scientists applying for a patent are benefiting from the labor of a much larger group or culture. In both the promotion of scientific progress and giving inventors their due, the Patent Office fails to fulfill its constitutional mandate.

HOW WE GOT HERE

The current policy of awarding patents on life emerged by default. In a sense, the policy was created through lack of decision. At first, through *Diamond v. Chakrabarty*, the Patent Office was led by the Supreme Court, but only in a legal sense. The Court said a genetically modified organism was patentable under the requirements of the Patent Office, but the justices left it to Congress to decide whether GMOs—and all life forms—*should* be patentable. Judge Berger, who presided over *Chakrabarty*, hinted at the need for a closer look by Congress at the negative implications of certain patents in his ruling. As he saw it, the Court's job was merely to decide if *Chakrabarty* met the utility requirement of a patent, not if, ultimately, genetically altered life forms should be patentable. Congress could amend section 101 on utility, Berger wrote, "so as to exclude from patent protection organisms produced by genetic engineering." Berger understood that Congress could do so under the same statute that does not allow nuclear weapons to be patented.[15]

In other words, the Supreme Court decision was not a decision over whether patents on life forms and genes should be allowed. It is as if you tried to patent a nuclear weapon. Although it has never been attempted, it is quite possible, under the Supreme Court's reasoning, that a nuclear weapon is patentable because it meets the Patent Office's three requirements of novel, non-obvious, and having real world utility. But the real question is not whether a nuclear weapon meets the requirements of the Patent Office; it is whether a nuclear weapon *should* be allowed to be patented. And, of course, Congress has passed a law that states that nuclear weapons cannot be patented, regardless of their "patentability." Congress has yet to consider the question of whether patents on life forms and genes should be allowed, despite the Supreme Court's suggestion.

As well, the Supreme Court had no idea of the consequential floodgates that would open because of that decision. The Court did not know that as a result of its saying a microorganism was patentable, the Patent Office would soon be granting patents on swaths of DNA found in our body, plants found throughout the world, and whole generations of mammals. The Patent Office has since made decisions on its own, expanding intellectual property rights to the point where it now grants patents that cover human clones. In 2002, a Washington think tank, the International Center for Technology Assessment (ICTA), discovered a patent that clearly covered the cloning of all mammals, *including humans*.[16]

A few congressional leaders have asked that the policy of granting life patents be looked at more carefully, but little has come of these requests. And the Patent Office does not allow input from any outside parties—including the public—when it considers whether to grant a patent. The Patent Office has received no guidance from Congress and the taxpaying public in over twenty years of granting life patents.

FREEDOM OF SPEECH AND
THE DEMOCRATIC PURSUIT OF VALUES

In the debate surrounding patents on life, there is a continual tension between the values assumed by the various and often conflicting points of view. For some, human beings, animals, plants, microorganisms, and all their parts are considered another form of property, to be bought and sold like a parcel of land. This is the attitude of the United States Patent and Trademark Office: "Genes are basically chemicals," Todd Dickenson, the former director of the Patent Office, told Congress.[17] For others, this material is valued as knowledge or information that should be shared for scientific investigation. The College of American Pathologists put it succinctly in its position paper against gene patents: "Information derived from mapping the human genome represents a naturally occurring, fundamental level of knowledge, which is not invented by man and should not be patented."[18]

For many indigenous groups, the plants and genes being patented represent cultural artifacts that cannot be claimed as inventions or discoveries. In many cases, it is not simply a plant or a drop of blood that is being taken and its biological components claimed as "invented"; it is also an important symbol of a culture's history and heritage. Foodstuffs, medicinal plants, and an individual's blood may be considered sacred and central to a culture's religious sensibilities.

"Some indigenous people say the patenting of this plant is the equivalent of somebody in their group patenting the Christian cross," said David Rothschild of the Coalition for Amazonian Peoples and Their Environment. Rothschild was referring to the patent awarded to a U.S. researcher in 1986 on an Amazonian plant called ayahuasca.

For still others, human beings, animals, plants, microorganisms, and all their parts can never be valued as inventions or property, but only as elements of nature. They believe such patents give ownership over natural processes or of life itself. Thus, Key Dismukes, study director for the Committee on Vision of the National Academy of Sciences, addresses the first "patent on life" on the Chakrabarty bacterium with this valorization:

> Let us get one thing straight: Ananda Chakrabarty did not create a new form of life; he merely intervened in the normal processes by which strains of bacteria exchange genetic information, to produce a new strain with an altered metabolic pattern. . . . The argument that the bacterium is Chakrabarty's handiwork and not nature's wildly exaggerates human power and displays the same hubris and ignorance of biology that have had such devastating impact on the ecology of our planet.[19]

What value one puts on biological material will determine one's judgment on the policy of awarding patents that cover molecular biology and living things. In turn, how one values such things is often indicated by what kind of investment—economic, spiritual, or relational—one has in such material. A family that has contributed the cells of its child to genetic research will place a different value on that genetic material than a company that will improve its stock after the announcement of the patent award on the same material.[20] Given such a range of values, particularly in the diverse ethnic, religious, and cultural milieu of the United States, it is difficult to imagine how any one policy on patents on life could ever be derived.

The current policy of granting patents on life was arrived at with no real public forum for exploring these diverse values, however. Without the advantages of hearing these differing views—without the advantages and right to free speech—the policy fails democratically.

Recent discussions regarding free speech involve the media or individuals whose art forms conflict with certain cultural norms, but a glance at the legal history of the right to free speech illuminates another aspect: the right for society to have access to a multitude of ideas and to hear from differing value systems. In order to arrive at a democratic value that will drive governmental policy, all values in a society must be weighed and heard from. This idea—that a democratic society profits from hearing many different ideas and differing sets of values and thereby arrives at just democratic policies—derived its reasoning from science itself. Oliver Wendell Holmes first took the scientific method and applied it to justice and democracy as a whole with his theory of a "marketplace of ideas":

the best test of truth is the power of that thought to get itself accepted in the competition of the market, and that truth is the only ground upon which [men's] wishes safely can be carried out. That at any rate is the theory of our Constitution. It is an experiment, as all life is an experiment.[21]

It is not that one policy will be arrived at which will please all parties, or that a compromise will be achieved in which all voices and values are equally respected. Rather, policy will be established after these voices have been heard and have been compared, tested, and analyzed in the "free market of ideas." In a democracy, it is not necessary that policy be directly decided by these voices, merely that these voices be heard.

As it stands now, patent policy restricts that debate to discussions between the Patent Office and the lawyers of universities and corporations and how they value human beings, animals, plants, microorganisms, and all their parts. This is a valorization which the universities and companies see as just, but it is not a democratic valorization.

It has been suggested that a more democratic forum could be created in the form of a regulatory body—similar to the Food and Drug Administration (FDA) or Federal Aviation Agency (FAA)—that would control and create policy on such biological "inventions," judging each within its specific context. The FAA is an especially apt model as it was developed from a need to balance patent protection with the dissemination of upstream knowledge in order for aerospace technology to progress.

Regardless of the form such a body ultimately takes, it is important that it be guided by the right to free speech, and that the ultimate value given to such "inventions" is reached in as democratic a manner as possible. If exclusive property rights are deemed appropriate to certain of these biological "inventions," then the regulatory body must further be guided by the constitutional mandate that these rights both promote science and give due credit to the actual "inventor." What is of paramount importance is that such a regulatory body be continually open to public comment, not just from scientists, universities, and private companies, but from all citizens, whose values must be considered in order to arrive at a democratically appropriate policy.

NOTES

1. Silvo A. Bedini, *Thomas Jefferson: Statesman of Science* (New York: Macmillan, 1990), 209; and William Eleroy Curtis, *The True Thomas Jefferson* (Philadelphia: J. B. Lippincott, 1901), 374–82.

2. "[A]s soon as I can speak of [the hemp-break's] effect with certainty," Jefferson wrote about one of his inventions, "I shall describe it anonymously in the public papers,

in order to forestall the prevention of its use by some interloping patentee." Curtis, *The True Thomas Jefferson*, 381.

3. "[A]s we enjoy great advantages from the inventions of others," Franklin wrote, "we should be glad of an opportunity to serve others by any invention of ours; and this we should do freely and generously." Benjamin Franklin, *The Autobiography and Other Writings* (New York: Penguin Books, 1986), 130.

4. See Bill Robie, *For the Greatest Achievement: A History of the Aero Club of America and the National Aeronautic Association* (Washington, DC: Smithsonian Institution Press, 1993), chapters 6, 8, and 11.

5. "Patent History," BountyQuest website, retrieved from www.bountyquest.com/patent/patenthistory.htm on February 8, 2002.

6. U.K. Commission on Intellectual Property Rights, *Integrating Intellectual Property Rights and Developmental Policy: Report of the Commission on Intellectual Property Rights* (London, England: Commission on Intellectual Property Rights, September 2002).

7. "Patently Problematic," *Economist*, September 14, 2002, 75.

8. Eric G. Campbell et al., "Data Withholding in Academic Genetics," *Journal of the American Medical Association* 287 (2002).

9. David Blumenthal, "Withholding Research Results in Academic Life Science: Evidence from a National Survey of Faculty," *Journal of the American Medical Association* 277 (1997).

10. "Climbing the Helical Staircase," *Economist*, March 27, 2003.

11. "Drugs Ex Machina," *Economist*, September 22, 2001. See also Peter Tollman et al., *A Revolution in R&D: How Genomics and Genetics Are Transforming the Biopharmaceutical Industry* (Boston: Boston Consulting Group, November 2001).

12. Deborah Smith, "Who Owns Your DNA?" *Sidney Morning Herald*, March 14, 2001. Bruce was part of a consortium of European churches that opposed the BRCA1 patent.

13. See Eliot Marshall, "Patent on HIV Receptor Provokes an Outcry," *Science* 287 (February 25, 2000): 1375.

14. Eliot Marshall, "HIV Experts vs. Sequencers in Patent Race," *Science* 275 (February 28, 1997).

15. *Diamond v. Chakrabarty* 447 U.S. 318 (1980).

16. "CBHD Denounces Patenting of Human Cloning," Center for Bioethics and Human Dignity Home Page, retrieved from www.cbhd.org/media/pr/pr2002-05-16.htm.

17. *2000 Gene Patents and Other Genomic Inventions*, hearing before the Subcommittee on Courts and Intellectual Property of the Committee on the Judiciary House of Representatives, 106th Congress, 2nd session, July 13, 2000, serial no. 121, 19.

18. Hearing before the Subcommittee on Courts and Intellectual Property, serial no. 121, appendix, 157.

19. Cited in Vandana Shiva, "North-South Conflicts in Intellectual Property Rights," *Peace Review* 12, no. 4 (2000): 504.

20. This is the context of the lawsuit brought against Miami Children's Hospital by the families of children affected by Canavan disease. Peter Gorner, "Parents Suing over

Patenting of Genetic Test," *Chicago Tribune*, November 19, 2000; and Eliot Marshall, "Families Sue Hospital, Scientist for Control of Canavan Gene," *Science* 290 (November 10, 2000): 1062.

21. *Abrams v. United States*, 630–31, quoted in Louis Menand, *The Metaphysical Club* (New York: Farrar, Straus and Giroux, 2001), 430.

• 4 •

New Enclosures: Why Civil Society and Governments Should Look Beyond Life Patents

Hope Shand

The right to live in a world in which living organisms cannot be patented is a paramount value. But debates over the ownership and control of new technologies must keep pace with emerging trends in science and the concentration in corporate power. This chapter argues that civil society and governments must adopt a broader formulation of rights and resistance, beyond "no patents on life," to guard against post-patent monopolies that threaten to erode diversity, democracy, and human rights.

Over the past two decades, intellectual property (IP) has become a powerful and controversial legal tool to enhance corporate monopoly and protect market share. IP has been a major factor in the growth and consolidation of the biotechnology industry. In the 1980s, the U.S. government took a series of steps to accommodate the corporate desire to patent life by redefining laws to allow for exclusive monopoly patents on all biological products and processes.

Since the 1980s, a growing number of civil society organizations (CSOs) and some governments have denounced life patenting as technically invalid and fundamentally inequitable. CSO critics contend that monopoly control over plants, animals, and other life forms jeopardizes world food security, undermines conservation and use of biological diversity, and threatens to increase the economic insecurity of farming communities. Instead of promoting innovation, patents are stifling research, limiting competition, and thwarting new discoveries.

Some industrial corporations are also becoming disenchanted with intellectual property—although for different reasons. The complexity and costs of patents are becoming problematic. Intellectual property laws are also becoming politically unpredictable. It is in this context that industry is seeking alternative mechanisms—"new enclosures"—to secure corporate control over biotechnology and other emerging technologies. After two decades of consolidation, five multinational corporations dominate the field of agricultural biotechnology. Patents become less relevant in oligopolistic markets and when other tools of monopoly are potentially cheaper and more far-reaching.

This chapter examines new enclosures and illustrates how they will supplement or replace IP as a means of strengthening corporate dominance over new technologies, and how they threaten democracy and dissent. It also examines the emerging field of nanotechnology and the need for civil society to broaden advocacy campaigns beyond "no patents on life."

WHY NEW ENCLOSURES?

Intellectual property is becoming increasingly costly. The application of patent law to living materials has resulted in immense and costly legal battles between corporations that are competing for ownership of strategic genes, traits, and biological processes. The transaction costs are enormous:

- The legal cost of obtaining a patent is approximately $10,000 in the United States, and it typically costs $1.5 million per side to litigate a patent.[1]
- In 2000, U.S.-based companies alone spent $4 billion on patent litigation.
- Start-up biotech companies are reported to be budgeting as much for patent litigation as they are for research expenditures.

Intellectual property is politically unpredictable. Compounding the uncertainties and costs associated with intellectual property, industry is now discovering that the "patenting of life" is politically contentious. There is growing public awareness accompanied by high-profile efforts to question, reform, and resist the patenting of life. Industry is worried that mounting political opposition to patents could lead to legislative changes that threaten its IP. For example:

- The 1999 UN *Human Development Report* states that "the relentless march of intellectual property rights needs to be stopped and questioned."[2]
- In August 2000 the United Nations Sub-Commission for the Protection of Human Rights recognized that the World Trade Organization's Trade-Related Intellectual Property Agreement could infringe on the rights of poor people and their access to both seeds and pharmaceuticals.[3]

- A report published in June 2002 by the United Kingdom's Nuffield Council on Bioethics argues that the vast majority of DNA sequences should not be eligible for patenting because they do not involve an inventive step.[4]
- The United Kingdom's Commission on Intellectual Property Rights concluded in September 2002 that intellectual property imposes costs on most developing countries and does not help to reduce poverty.[5] Because of the restrictions patents place on seed-saving farmers and researchers, the commission recommends that developing countries should generally not provide patent protection for plants and animals.[6]
- In its 2002 report *Genomics and World Health*, the World Health Organization concludes: "The current position regarding DNA patenting is retarding rather than stimulating both scientific and economic progress. The monopolies awarded by patents on genes as novel chemicals, therefore, are not in the public interest."[7]

The political and practical uncertainties of intellectual property are increasingly unacceptable to industry, and this is propelling industry to seek new control strategies—new enclosures. Three major categories of new enclosures (in this case, all relating to agricultural technologies) are identified below:

1. Biological monopolies on germplasm.
2. Remote sensing and other technologies to detect or trace living organisms.
3. Legal contracts.

BIOLOGICAL MONOPOLIES

The best-known examples of new enclosure mechanisms are the controversial genetic use restriction technologies (GURTs), better known as Terminator and Traitor technologies. GURTs involve the use of genetic switches, triggered by the application of external chemicals, to control a plant's genetic traits. *Terminator* refers to plants that are genetically modified to switch on or off the trait for seed sterility. Seeds harvested from Terminator crops will not germinate if replanted the following season. The technology aims to prevent farmers from saving seed from their harvest, thus forcing them to return to the commercial seed market every year.[8] Patents are a legal mechanism to prevent farmers from saving and replanting proprietary seed. If commercialized, Terminator seeds would offer a biological mechanism to eliminate farmer seed-saving. For corporate gene giants, genetic seed sterilization offers a stronger and more far-reaching monopoly than intellectual property because, unlike patents, Terminator technology would not be time-limited; it would offer no exemption for researchers, no provision for compulsory licensing, and no need for lawyers.

REMOTE SENSING AND SURVEILLANCE

Earth-observation satellites—operating beyond the limits of national sovereignty—are already being used by governments, civil society, and industry to collect images and geospatial information on human activities and the natural environment. The first satellites were government owned and used strictly for military purposes. The world's first commercial earth-observation satellite, launched in September 1999, provides images with better than one-meter resolution. Today, new constellations of smaller, cheaper, and more agile commercial satellites are being financed, built, and operated by private firms that hope to profit from satellite imagery, geospatial information products, and related services.

Satellite imagery and geospatial information technologies have potential to promote transparency[9] and benefit agriculture, but they also threaten to diminish the rights of farmers and farm communities. Remote sensing and biodetectors are being used by corporations and governments to: (1) enforce proprietary rights and regulatory or contract compliance, and (2) identify, monitor, and control germplasm, territory, and labor.

The following examples illustrate potential uses of remote sensing that could threaten civil liberties and farmers' rights:

- Florida's Department of Agriculture is proposing to use satellite photos to spy on its biggest citrus competitor, the Brazilian state of São Paulo. A Florida-based trade association told the *Wall Street Journal*, "Brazil's oligopoly controls world orange juice prices . . . to the detriment of the Florida industry."[10] Satellite photos of Brazil's citrus crop will cost an estimated US$1 million to $1.5 million, a price that is out of reach for the majority of agricultural producers.
- In 2000, the National Seed Institute (INASE) of Argentina proposed to use satellite surveillance to stop illegal seed commerce. "The current law permits a producer to save seed for his own use, but in no way is he authorized to sell the seed, or trade it for other inputs or machinery and he can't even give it away to other producers, because that act of delivery is prohibited and, in the end, will be prosecuted," warned INASE officials.[11]
- In early 2001, the Argentine government announced that it would use satellite imagery to monitor farmers' crops in an effort to halt tax evasion.[12]

LEGAL CONTRACTS

Legal contracts are being used as new enclosure mechanisms to control germplasm, technology, and research. Increasingly, the seed industry provides proprietary seed to farmers under contractual agreements that prohibit the farmer from saving, reusing, or selling any of the harvested crop as seed. With the commercialization of genetically modified (GM) seed, contract provisions known as "technology user" or licensing agreements are commonplace—and controversial. Sometimes contract agreements are used in combination with intellectual property restrictions; other times they are used alone.[13]

Seed industry giants such as Pioneer (DuPont) and Monsanto routinely use technology user agreements when they sell genetically modified seed. The contracts not only restrict the use of harvested seed but go far beyond intellectual property by dictating conditions for using seed and related inputs and establishing limits for liability and legal recourse, and even conditions for post-harvest marketing. Consider the following examples:

- *Liability Limits:* Farmers who sign Monsanto's 2001 technology agreement must accept the company's Exclusive Limited Warranty, which severely limits Monsanto's liability for any and all losses, injury, or damages that result from the use or handling of a product containing Monsanto's gene technology.[14]
- *Right of Venue:* Right of venue clauses allow the seed company to force breach of contract disputes arising from technology agreements to be settled exclusively in court jurisdictions that are generally more favorable to the corporation, and typically make defense against infringement charges more costly to the farmer.[15]
- *Dictated Farming Conditions:* According to economist Dwight Aakre of North Dakota State University, Monsanto's 2001 technology agreement for RoundUp Ready GM crops states that the producer has responsibility for crop isolation to insure that pollen from GM crops does not trespass on a neighbor's crop. Aakre believes that growers of GM crops are exposing themselves to potentially huge financial risks by signing gene technology agreements.[16]
- *Post-Harvest Liability:* A farmer who signs Pioneer's contract for both YieldGard and LibertyLink gene technology "agrees to keep the harvested grain from these hybrids out of European grain export channels."[17] Monsanto's 2001 agreement on RoundUp Ready crops has similar provisions. Dwight Aakre warns farmers: "Signing that agreement means you accept a risk that you have very little control over. If a ship load of grain arrives at one of these export markets, is tested and

found to contain unapproved genetics and the source can be traced back to your farm, what is your responsibility?"[18]

In North America, Monsanto has aggressively monitored and prosecuted seed-saving farmers with the help of private investigators, dubbed "gene police" by the *Washington Post*.[19] The company has filed more than 475 lawsuits against farmers for patent infringement and violation of technology user agreements (the exact number is not known).[20] Given the fact that Monsanto's genetically modified seed technology accounted for over 90 percent of the total world area planted in GM seed in 2001, the potential impact of contract agreements extends far beyond North America.[21]

BEYOND LIFE PATENTING: NANO-SCALE TECHNOLOGIES ARE NO SMALL MATTER

"Nanotechnology is new, and this means a lot of possible patent protection could be very broad," states Michael Hostetler, an intellectual property lawyer. According to Hostetler, "This is a time of great opportunity; nanotechnology is such a wide-open field right now. It's rather like a land grab in the Wild West."[22] With the emergence of powerful new technologies at the nano-scale, the Genetic Bill of Rights and other civil society campaigns against the patenting of life must be broadened to confront the realities of new technologies and new control strategies.

Today, the capacity of scientists to manipulate matter is taking a giant step *down*—from genes to atoms.[23] *Nanotechnology* refers to the manipulation of atoms and molecules to make new products. At the nano-scale, where objects are measured in billionths of meters, the distinction between living and non-living things is blurred. The raw materials for nanotechnology are the chemical elements of the periodic table, the building blocks of all matter—both living and non-living. Taking advantage of quantum physics, nanotech companies are already engineering novel materials that may have entirely new properties never before identified in nature. Worldwide, government and industry will spend an estimated $8.6 billion on nanoscale science and technology in 2004; virtually all Fortune 500 companies are investing in nanotech research and development.[24]

Atomic-level manufacturing provides new opportunities for sweeping monopoly control over both animate and inanimate matter.[25] Will nanotechnology follow biotech's passion for sweeping patent claims that could give a handful of corporate giants exclusive monopoly over the products and processes that are the raw materials for a new economy of manufacture? With the emergence of nano-scale technologies will we see sweeping patent claims on chemical elements?

Will it be possible to patent new elements? In essence, patenting at the nano-scale could mean monopolizing the basic elements that make life possible.

CONCLUSION

Intellectual property is not the only mechanism being used by corporations to achieve market monopolies and long-term control over new technologies. New enclosures threaten to erode the rights of farmers and workers, undermine national sovereignty, and promote corporate consolidation. A broader formulation of rights and resistance, beyond Article 2 of the Genetic Bill of Rights, is needed to guard against post-patent monopolies. Efforts to resist and reform intellectual property must not be limited to campaigns against the patenting of life. Nano-technology, the science of manipulating matter at the level of atoms and molecules, is positioning the world's largest companies to seek patent claims on the building blocks of the entire natural world.

In order to safeguard human rights and democratic dissent, new enclosures must be carefully monitored, analyzed, and independently regulated. Action is needed at all levels—from local communities and national governments to intergovernmental bodies. Because transnational enterprises and technologies operate beyond the boundaries of any single country, reform will require debate, oversight, and monitoring at the United Nations level. In 1974 the United Nations established the Centre on Transnational Corporations—but its programs withered and the Centre closed in 1993. While it remains vitally important to resist and reform intellectual property monopolies, the international community must regain the capacity to monitor and regulate the activities of transnational enterprises. Beyond corporate governance, the international community must also create a new body with the mandate to evaluate and accept or reject new technologies and their products through an International Convention on the Evaluation of New Technologies. In response to a flood of patent applications involving nanoscale products and processes, in October 2004, the U.S. Patent and Trademark Office established a new registration category for nanotechnology inventions.

NOTES

1. John Barton, "Reforming the Patent System," *Science* 287 (March 17, 2000): 1933–4.

2. United Nations Development Program, *Human Development Report 1999* (New York: Oxford University Press, July 1999).

3. United Nations, Sub-Commission on Intellectual Property Rights and Human Rights, Commission on Human Rights, "Resolution on Intellectual Property Rights and Human Rights," E/CN.4/Sub.2/2000/7 (August 17, 2000).

4. Emma Dorey, "Nuffield Slams DNA Patents," *Nature Biotechnology* 20 (September 2002): 864.

5. Commission on Intellectual Property Rights, "Independent Commission Finds Intellectual Property Rights Impose Costs on Most Developing Countries and Do Not Help to Reduce Poverty," press release, September 12, 2002, www.iprcommission.org.

6. Commission on Intellectual Property Rights, *Executive Summary: Integrating Intellectual Property Rights and Development Policy* (London: CIPR, September 2002), 17.

7. Advisory Committee on Health Research, *Genomics and World Health* (Geneva: WHO, 2002), 19.

8. For further information, visit www.etcgroup.org, the website of the ETC Group.

9. John C. Baker, Kevin M. O'Connell, and Ray A. Williamson, eds., *Commercial Observation Satellites: At the Leading Edge of Global Transparency* (Santa Monica, CA: RAND and the American Society of Photogrammetry and Remote Sensing, 2001).

10. B. McKay and M. Jordan, "Orange Juice Rivalry Spurs Florida Plan to Spy on Brazil," *Wall Street Journal*, January 28, 2002.

11. "Tecnología satelital para detectar comercio illegal," *Revista Chacra* (2000), www.revistachacra.com.ar/notas/cne200006n06.htm.

12. "Satellite Photos to Aid Efforts Against Tax Evasion in Argentina," *Financial Times Global News Wire*, January 9, 2001.

13. Neal D. Hamilton, "Possible Effects of Recent Developments in Plant Related Intellectual Property Rights in the U.S.," prepared for presentation at the International Seminar on Effects of Intellectual Property Rights on Agriculture in Developing Countries, March 7–8, 1995, Santa Fe de Bogota, Colombia. Neil Hamilton, director of Drake University Agricultural Law Center, notes, for example, that Stine Seed Company (a soybean seed company based in the United States) has relied on contract provisions rather than intellectual property to enforce ownership of its soybean seeds. If a farmer or breeder violates the purchasing agreement by using or selling a harvested crop for seed or breeding purposes, the company is authorized to use breach of contract claims in local courts. The company's use of contract provisions limits the rights that farmers and breeders would normally have if the seed was protected under standard plant variety protection laws.

14. Eva Ann Dorris, "Monsanto Contracts: To Sign or Not to Sign," *Mississippi Farmer*, December 1, 2000.

15. Brief *Amici Curiae* of American Corn Growers Association and National Farmers Union in Support of the Petitioners, *J.E.M. AG Supply, Inc. v. Pioneer Hi-Bred International, Inc.* (submitted by Joseph Mendelson III and Andrew C. Kimbrell, International Center for Technology Assessment, no. 99-1996, 2001).

16. North Dakota State University Agriculture Communication, news release, "GMOs Bring Increased Liability Risk for Producers," March 22, 2001.

17. Pioneer Hi-Bred, *YieldGard Product Use Guide* (2001).

18. North Dakota State University Agriculture Communication, news release, March 22, 2001.

19. Rick Weiss, "Monsanto's Gene Police Raise Alarm on Farmers' Rights, Rural Tradition," *Washington Post*, February 3, 1999, 1.

20. The statistic is cited in *Amici Curiae* brief prepared by Joseph Mendelson and Andrew C. Kimbrell.

21. C. James, "Global Review of Commercialized Transgenic Crops: 2001," *ISAAA Briefs* 24 (2001). According to ISAAA, the global area devoted to transgenic or genetically modified (GM) crops increased more than thirty-fold, from 1.7 million hectares in 1996 to 52.6 million hectares in 2001. According to Monsanto's website, the company's GM seed technology covered 48 million hectares (118 million acres) in 2001.

22. Sandra Helsel, "Shootout at the IP Corral," *Nano Circuit*, October 4, 2002. Available at www.nanoelectronicsplanet.com/nanochannels/circuit/article/0,4028,10501_1476651,00.html.

23. ETC Group, *The Big Down: Atomtech—Technologies Converging at the Nano-scale* (January 2003). Available at www.etcgroup.org.

24. ETC Group, "Oligopoly, Inc.," *ETC Communiqué*, November/December 2003. Available at www.etcgroup.org/article.asp?newsid=420.

25. ETC Group, "Patenting Elements of Nature," *Genotype*, March 25, 2002. Available at www.etcgroup.org/article.asp?newsid=308.

Life Patents Undermine the Exchange of Technology and Scientific Ideas

Jonathan King and Doreen Stabinsky

The breakthroughs in molecular genetics, biochemistry, and cell biology that gave birth to the biotechnology revolution are a defining aspect of the beginning of the twenty-first century. The resulting increases in biological understanding offer extraordinary new possibilities for preventing and treating disease, for a much deeper understanding of the interactions of organisms with each other and their environment, and for entirely new manufacturing technologies. Intertwined with the development of biotechnology has been its rapid commercialization, initially for pharmaceutical and agricultural production.

An unforeseen and deeply troubling aspect of this commercialization is the transformation of biological entities that are the products of hundreds of millions of years of evolution into private property. This is taking place through the radical extension of patent law to encompass gene sequences and subsequences, cell lines, and genetically modified organisms, including plants and animals. These profound changes are being carried out through the administrative procedures of the U.S. Patent and Trademark Office (PTO), without public debate or congressional oversight.

Myriad Genetics owns patents on two mutated genes (BRCA1 and BRCA2) linked to hereditary breast cancer. Incyte Genomics and Celera have filed over 10,000 patent claims on human gene fragments, many of unknown biological function. Human Genome Sciences has patented the entire genomes of important bacterial pathogens affecting public health. Monsanto owns the patent rights to more than 90 percent of all commercially available genetically modified crop varieties. Rice-Tec Corporation has obtained a patent on basmati rice, grown in India for thousands of years.

Such life patents represent a radical departure from the cultural traditions of human societies. Farmers have always owned the crops they grew, but they had no rights over the same crops grown by others, nor any legal ability to restrict others from growing those crops. The Monsanto patents on transgenic cotton, soy, canola, and corn extend to all the progeny of such plants and allow Monsanto to prevent farmers from saving the seed of crops they have grown and planting them the next season. The cloning of Dolly was not announced until the Roslin Institute had filed patents not just for cloned sheep, but for all animals produced anywhere in the world by a similar process. Such private expropriation of fundamental biological resources reflects a qualitative change in access to basic biological knowledge and to the relations between human society and the natural world.

A patent allows the owner to exclude others from using or benefiting from the patented invention, process, or construct of matter. U.S. patent law, by granting a monopoly for twenty years to the patent holders, allows a company to prevent other efforts to produce or utilize the "invention," even if for medical purposes or human welfare. In the battles between corporations, this takes the form of infringement suits, injunctions against sales of products, and other forms of legal intervention. Hundreds of millions of dollars' worth of such suits are brought regularly as corporations maneuver for control of emerging markets.

U.S. patent laws were originally written by Thomas Jefferson. Jefferson was an active plant breeder and corresponded with leading breeders in Europe. Nonetheless, the patent laws as developed by Jefferson excluded animals and plants from their coverage. Jefferson was clear that patents were a form of monopoly. He believed the role of patents lay not in the generalized protection of private property, but in the limited and specific purpose of ensuring that creative and inventive individuals were able to make a living from their work, and thus continue to contribute to society. He wrote that whenever this monopoly was contrary to the public interest, the public interest would take precedence.

With the commercialization of plant breeding and seed production in the 1920s, breeders attempted to restrict competition through inclusion of ornamental plants and other hybrids under the patent laws. Resistance from consumer groups and farmers prevented this, but breeders were granted some "protection" by the Congress through the passage of separate legislation, namely the Plant Patent Act of 1930 and the Plant Variety Protection Act of 1970. Organisms in general and their genes, proteins, or component cell lines remained excluded from general patent monopolies.

This two-hundred-year-old legacy was breached in 1980 through the Supreme Court decision in *Diamond v. Chakrabarty* that upheld granting a patent for a genetically engineered bacterium.[1] The 5–4 decision was narrowly constructed with respect to genetically modified microorganisms. In the years following, with lobbying from the pharmaceutical, biotech, and agrobiotech interests, the U.S. PTO began issuing patents on genes, human cell lines, and plant strains. Patent offices in some other nations followed suit, and a 2001 decision by the European Court of Justice compelled member states to permit the patenting of plants, animals, and their component parts.[2] U.S. corporate interests have used the vehicle of the World Trade Organization (WTO) and its Agreement on Trade Related Aspects of Intellectual Property Rights (TRIPS) to aggressively press for global enforcement of life patent protections.

The traditional criteria for the granting of patents are novelty, utility, and non-obviousness. Prior to the *Chakrabarty* decision, patents were limited to true inventions of machines, novel processes, synthetic materials, and related "compositions of matter." Thus minerals cannot be patented because they are found or discovered, not invented. Such "products of nature" have historically been excluded from patent protection. As biologists, we are appalled at the claims that the determination of the nucleotide sequence of a gene represents a novel invention. However, the underlying issues are questions of social policy and not legal interpretation. Patent laws are passed, modified, and abrogated by the U.S. Congress just as other laws governing the country. The Constitution simply states that the Congress shall have the power to grant patents "to promote the Progress of Science and useful Arts."

We outline below some ways in which life patents threaten our ability to reap the fruits of molecular genetics, biochemistry, cell biology, and other scientific and technological advances of the biotechnology revolution.

RESEARCH AND SCHOLARSHIP

The biotechnology revolution in the United States was the product of a broad-based biomedical research and training enterprise centered at colleges, universities, and medical schools throughout the nation. Essential to these efforts was the free communication and exchange of materials and ideas, and the organization of research in the public interest. Major scientific advances, such as the determination of the amino acid sequences that make up protein chains, were openly communicated and entered the public domain. The enormous inventiveness during this period occurred without patents.

Patent law requires that the subject of the patent has not been revealed as "prior art." Oral reports, abstracts, grant proposals, and published papers all constitute prior art. Thus individuals or groups planning to file for a patent have to avoid public disclosure of the work prior to the filing of the patent claim. Patent attorneys regularly advise researchers to restrict discussions with colleagues and hold back reporting results, so as not to jeopardize planned patent submissions. Indeed, academic scientists cite efforts to protect proprietary interests as a leading cause of data withholding and non-disclosure of research findings.[3]

Since 1980, the number of patents claimed by universities has risen dramatically, due in large part to laws enabling academic scientists to retain ownership rights to federally funded inventions.[4] Entrepreneurship and commercialization increasingly compete with the traditional norms of science. The resulting undermining and reversal of the academic culture of open communication and exchange is one of the most destructive impacts of life patents.

HEALTH AND MEDICAL CARE

In the health care field, patents not only retard progress at the level of research but interfere with the delivery of health care. The key commercial value of a patent is the ability to prevent competitors from developing or delivering a related (or superior) process or product; this holds for public and non-profit institutions as well as other companies.

Thus when the Biocyte Corporation obtained a patent on the use of blood cells (obtained from the umbilical cord) for a variety of therapeutic situations, health care professionals responded: "We . . . join in protesting . . . the granting of this patent to Biocyte Corporation. We are also concerned that this patent may discourage and threaten activities of non-profit cord blood banks, clinicians, patients, parents and the volunteers who support them."[5]

Indeed, empirical data confirm that patents impede access to key biomedical technologies and services. In a recent U.S. survey of laboratories offering clinical genetic testing, 25 percent of respondents reported having stopped offering one or more tests because of a pre-existing patent or license, while 53 percent reported having decided not to develop a new genetic test for the same reason.[6] These include tests for genes associated with breast cancer, cystic fibrosis, hemochromatosis, and a host of other diseases. It is noteworthy that in India, Brazil, and other countries, patent laws exclude pharmaceutical and other health care products from being patented on the grounds that these patents could compromise public welfare.

FOOD AND AGRICULTURE

Intellectual property rights on seeds deprive farmers of free access to an essential public good: crop seed. Farmers for millennia have saved the seed of their harvests for replanting the following year; they continue to do so even in the industrialized agricultural systems of the United States. Patents on seeds are a legal means to make sure farmers buy seed every year; in the last few years, Monsanto has taken hundreds of farmers to court for planting saved seed from their genetically engineered varieties.[7]

The problems of intellectual property protection on seed become magnified by the economic consolidation currently taking place in the seed industry. Monsanto and DuPont alone may now control half of the U.S. soy and corn seed markets. At the international level, the ownership situation is similar: in Argentina, Monsanto owns 40 percent of the corn seed market; in Brazil, 30 percent. Monopoly control in an oligopolistic market is a sure recipe for higher seed costs.

COMMERCIAL ACTIVITY

Corporate spokespersons claim that without patent protection, important technologies will not be developed. In fact, what patent protection ensures is not technological development, but suppression of competition. Patents are as often used to prevent the development of new technologies as to exploit them. Some of these barriers have been further explored by Michael Heller and Rebecca Eisenberg.[8] In the pharmaceutical and biotechnology industries, the role of patents is as a mechanism of monopoly pricing rather than technology development.

BRINGING THE ISSUES INTO THE DEMOCRATIC PROCESS

We conclude from all these arguments that the privatization of life forms through the extension of patent laws represents disastrous social policy. Critical decisions are being made by an out-of-control U.S. Patent and Trademark Office, operating far outside its constitutional mandate.

In Europe, Africa, Southeast Asia, and South America, significant social movements oppose life patents. Dramatic public demonstrations occurred in India in response to W. R. Grace's obtaining patents on the neem tree, and these were followed by a vigorous battle in India's upper house of parliament to resist the GATT intellectual property requirements. Recently, the African Group of member states to the WTO has called for an amendment to global intellectual property standards to prohibit patents on all life forms. The coalition argues that patents on plants, animals, and microorganisms "contravene the basic tenets on which patent laws are based: that substances and processes that exist in nature are a discovery and not an invention and thus are not patentable."[9]

The United States and other nations of the world have recognized the necessity of maintaining common resources such as the oceans, the atmosphere, and the moon outside national sovereignty or corporate property. The earth's life forms need to be given the same consideration. The first step is the open discussion of these questions, not in patent courts, but in colleges and universities, in professional societies, and in the U.S. Congress.

NOTES

1. *Diamond v. Chakrabarty*, 447 U.S. 303 (1980).

2. European Court of Justice Directive 98/44/EE.

3. Eric G. Campbell et al., "Data Withholding in Academic Genetics: Results from a Nationwide Survey," *Journal of the American Medical Association* 287, no. 15 (2002): 1939–40.

4. Lita Nelsen, "The Rise of Intellectual Property Protection in the American University," *Science* 279 (March 6, 1998): 1460–1.

5. Declan Butler, "U.S. Company Comes Under Fire over Patent on Umbilical Cord Cells," *Nature* 382 (July 11, 1996): 99.

6. Mildred K. Cho et al., "Effects of Patents and Licenses on the Provision of Clinical Genetic Testing Services," *Journal of Molecular Diagnostics* 5, no. 1 (2003): 3–7.

7. See, for example, *Monsanto Canada Inc. v. Schmeiser*, 2004 SCC 34 (Supreme Court of Canada, 2004).

8. Michael A. Heller and Rebecca S. Eisenberg, "Can Patents Deter Innovation? The Anti-Commons in Biomedical Research," *Science* 280 (May 1, 1998): 698–701.

9. GRAIN, "Africa Group Position on TRIPS," www.grain.org/bio-ipr/?id=27.

Part III

GENETICALLY ENGINEERED FOOD

"All people have the right to a food supply that has not been genetically engineered."

Article 3 of the Genetic Bill of Rights

· 6 ·

Food Free of Genetic Engineering: More Than a Right

Martha R. Herbert

*W*hat should we do when two conflicting assertions of rights are in whole or in part mutually exclusive? Can the "right to modify food genetically"—a right asserted mainly by producers of genetically modified foods and producer-friendly policy makers—coexist with the "right to have access to food that is not engineered"? This chapter argues that the new technology of genetic engineering should not be preemptive of traditional food. If genetically engineered (GE) food can displace and eliminate cultivation of food that has not been genetically engineered, then the insistence on access to food free of genetic engineering is at the same time a call to restrict GE food, and if necessary to curtail it severely or entirely.[1] The right to food free of genetic engineering simply cannot be compromised.

Objections to genetic engineering of food are fundamental. Given the deeply questionable premises of genetically modifying food, it is not surprising that the technology has not delivered on its promises.[2] The tables need to be turned. For reasons that I will spell out on multiple levels, we cannot rest on insisting that food free of genetic engineering be merely preserved as an option. Instead, we need to be arguing that agribusiness should not have the right to implement the genetic engineering of food at all, given the inherent unpredictability of the technology and its many risks, the poor prioritizing it represents, and everything valuable that it displaces.

CULTIVATION VS. PRODUCTION

To insist on access to non-GE food is a good start, but it is not sufficient for dealing with the broad ramifications of genetically modifying the food supply. The issues here go far beyond health and the testing and labeling of new food products, and very far beyond choice in the supermarket aisles. They reach into fundamental questions about how we evaluate technologies. Demonstrating that a technology appears to "work" is short sighted if the longer-term consequences and ripple effects of the technology are ignored. The issues also reach into questions about how we organize agriculture and how we keep ourselves and our fellow living beings alive. Industrial farming, of which GE food is only the most recent example, has forced a transition from food *cultivation* to food *production*.[3] The emphasis on *production* dismisses an enormous range of ecological and cultural considerations related to food.[4] Proponents of GE food promise that genetic engineering will increase food productivity. But they ignore a host of other relevant domains—including not only ecological and health concerns, but also the communities and cultures of farming, the cultural resonance of cuisine, and the historically contingent and problematic urban-rural split.

Even in its own narrow productivist terms, genetic engineering is likely to yield not productivity but its opposite—crop failures, diseases, or blights from unforeseen vulnerability of genetically manipulated strains cultivated as widespread monocultures. Moreover, a serious analysis of the causes of world hunger reveals that, for many social and economic reasons (including maldistribution of ample food stocks), productivity is not the issue.[5] Fundamentally, insistence that food be free of genetic engineering is a critically important issue because ecological and cultural sustainability are at stake. Science is now able to develop "gentle, thought-intensive technologies,"[6] to advance beyond the industrial and engineered monocultures, including insertion of gene sequences, that may be characterized as aggressive, energy and input laden, and hype intensive. What we need for both physical survival and for a future worth living is a scientifically sophisticated but context-sensitive and culturally rich recovery of *cultivation*. Genetic engineering of food and the vested interests that obstruct balanced debate about it are obstacles to this deeply needed advance.

KEEP THE DEBATE BROAD, FULL, AND TRANSPARENT

Keeping the food supply free of genetic engineering should be guaranteed not only as a right but as a necessity. This right and indeed necessity can be defended on many grounds—including views about molecular genetics, cell biology, plant and animal physiology, ecology, economics, health, culture, even aesthetics. GE food proponents often call for arbitration of controversy over this technology through what they call "sound science," which at present consists of hastily conducted, short-term studies of this technology. Such "studies" do not begin to carry sufficient weight to satisfy even scientific issues, let alone concerns at many other levels. Proponents sometimes attempt to restrict debate to health issues, and then to foreclose discussion on the grounds that studies to date show no health risks. But this gambit ignores, indeed denies, many other concerns beyond human health. Attempts to restrict the debate to human health and to ignore the many plausible scientific questions as well as other types of concerns are as objectionable as the technology itself. Both the technology and the controversy over it are so new and so many sided that any call for hurried approval of GE food, particularly without protecting the availability of food free of this kind of manipulation, can only result from ideological zeal or financial interest.

To its opponents, the genetic engineering of food is a technology based on limited and parochial assumptions,[7] deplorably naive about organisms, oblivious to ecology, economically motivated, and blind (at least in part deceitfully so) to the real causes of hunger in the world. Yet genetically engineered crops and animals have been rushed into large-scale production with inadequate scientific evaluation and public discussion. Why? The reasons are of two kinds: belief systems and economics.

Proponents of genetically engineered food have consistently resisted engaging opposing perspectives. Regulators welcome favorable assessments, even if they are of poor quality, but give critical assessments a hard time even if they meet rigorous standards and are peer reviewed.[8] In addition, while there has been abundant funding for genetic engineering research, little money is available for context-sensitive agroecological approaches.[9] One reason is that genetic engineering can easily lead to patentable products and the promise of profit, while agroecology, though more sustainable, generally cannot lead to such prospects of economic gain. These biases have been incorporated into national policy; for example, international trade legislation includes funds for promoting agricultural biotechnology but not for seriously assessing it or developing agroecological, non-engineered approaches.[10] Thus, it is important for the public to understand that we do not see "equal time" or the operation of "unbiased science" in allocation of research resources. Serious conflicts of interest have dogged government- and industry-sponsored inquiry, with commissions composed predominantly of members with industry ties and funding empowered to consider the merits of GE food.[11]

Proponents of GE foods do not appear willing to engage in open and transparent debate. We may attribute some of this to vested interests, but that does not fully explain the problem. Many GE food proponents not only fail to address the concerns of GE food critics but appear unable even to comprehend the criticisms. They frequently claim that they themselves (the proponents) are uniquely "scientific" and their critics merely "emotional." Sometimes this rhetorical strategy is a disingenuous public relations maneuver. But it also reflects genuine naivete. Arguments about GE food's threats to organismic, ecological, and cultural complexity and diversity may simply be incomprehensible to many GE enthusiasts. They appear to see molecular genetics as the definitive universal code of life, whose encompassing truth must override all prior frameworks. Because the DNA code translates into amino acids in fundamentally the same way across species, the particularities of species differences seem incidental. GE thus rests on a purported universalism that holds that particularities and local details have a lesser importance than the general abstractions that can be distilled out of them. It involves a belief (and only a belief, though its adherents seem to think it is "fact") that all analogical processes at every level, including organismic processes, lived experience, and more, can be digitized into code (like the genetic code) without loss of nuance.

One also sees an emboldened triumphalism, a sense of mission to improve the world on the basis of what are seen as "truths" revealed by molecular genetics. All human and other organic frailties are seen as susceptible to remediation by engineering or genetic recoding. Calls by doubters to consider problems that arise where recoding cannot be directly applied are considered naive and irritating distractions, willful obstacles to the "incredible potential" of genetic modification. Yet ironically this investment in the "universalism" of the genetic code has even interfered with genetic science itself, because while a growing number of studies have identified ways, sometimes species specific, that non-coding DNA as well as non-DNA proteins modulate gene expression—and moreover may do so in ways that differ among species—these findings cannot comfortably be incorporated into an ideological framework of genetic universalism and gene dominance.

GENETIC ENGINEERING AS TECHNOLOGICAL MESSIANISM

Technological messianism dovetails elegantly with the economic forces driving genetic modification. Inserting a specifically characterized gene sequence into an organism has been considered adequate justification for patenting the organism.[12] This patented seed offers numerous benefits to the patent-holding proprietor. It allows a new kind of ownership of organisms. Contractual relations can apply, allowing the revenue stream to be assured with new mechanisms not available to non-engineered varieties of agricultural organisms. This patenting and turning of living beings into intellectual property occurs in a market system where the goal is accumulation of profit. It is worth reflecting that getting rich means having more money, and "money" itself is an abstraction that is dissociated from the particular qualities of the commodities that are produced and sold. One can get rich from selling corn flakes or nerve gas—it doesn't really matter to the "bottom line." In this setting, the compulsion to implement a more efficient means of capital accumulation overwhelms all other considerations. Thus, the mission to improve the world by redesigning it according to "genetic universals" complements the economic drive to control the market—and the world—according to the "universal money abstraction." Both the money abstraction and the genetic abstraction are divorced from any commonsense reality checks because they are divorced from any particular loyalties to specific context, whether it be place, species, person, or culture. Those who pursue the "money abstraction" and the "genetic abstraction" are, in terms of the "logic" of their activities, impervious to arguments coming from any domain of particularity that is outside their frame of reference. Such particular concerns may simply not register in the mind of anyone staying within this abstract framework.

The technological messianism of GE food advocates thus coexists poorly with other belief systems. Certainly this inability to coexist with other frames of reference characterizes messianisms of many kinds. The problem is that the genetic engineering of food is more than a belief system; it is a technology—and moreover, a technology which utilizes living organisms as its substrate and transforms them in unprecedented ways. GE foods do not merely represent a belief system; they *embody* it. GE foods incorporate the belief system that conceived them in their very tissues, their very flesh, indeed their genes, in a manner that goes beyond previous breeding techniques of industrial agriculture. They thus do not assert themselves merely as ideology or dogma, but even more as material—and organismic—force. And as a material, living force enlisted in a messianic mission, they not only ideologically oppose but, even more, materially—and reproductively—displace non–genetically engineered organisms. Once an organism is genetically modified, there is no going back. And once genetically engineered organisms are in the environment, gene-sharing with non–genetically engineered wild species cannot be controlled.[13]

This aggressive, intrusive character of GE food is not due just to the nature of the technology and its ecological risks. It also appears to be an explicit market strategy. In the words of one industry spokesperson: "The hope of the industry is that over time the market is so flooded [with GE food] that there's nothing you can do about it. You just sort of surrender."[14] A U.S. government official said, without distress: "In four years, enough GE crops will have been planted in South Africa that the pollen will have contaminated the entire continent."[15] From this point of view, the biotech industry might privately perceive the genetic contamination of maize by transgenic DNA in its center of origin in Oaxaca, Mexico, as a welcome development.[16]

Thus, the perception of genetic engineering of food as an intrusive and self-propagating biological colonialism cannot be refuted by recourse to scientific studies finding no evidence of hazard, because hazard, while an issue, is not the *only* issue. But the dominant attempt to keep the discussion narrow, the dismissal of the value of many other levels of concern, and the aggressive attempts to restructure and gain control of global agriculture all reinforce the perception that GE food advocates are perpetrators of a new level of colonialism. The right to food free of genetic engineering—and indeed strong opposition to the right to pursue this questionable technology—are both thus critical bulwarks against being engulfed and devoured by an insensitive, greed-driven ideological monolith.

THE MULTILEVEL OBJECTIONS
TO GENETICALLY ENGINEERED FOOD

Critics of genetically engineered food have not shared the conversion experience of the enthusiasts. GE food proponents may allege that ignorance is the reason for the critics' failure—or more accurately, their *refusal*—to see the genetic code as a comprehensive universalism. But the reasons for rejecting GE food are substantive and span multiple levels, from molecular genetics all the way to ecology and culture. Insistence on the right to—and need for—food free of genetic engineering is grounded in all these levels. It is unlikely that all of these arguments could ever be refuted by GE food proponents, which is presumably why most of these levels are so often excluded from official, industry-influenced debate.

At the level of genetics, there is abundant evidence that the genetic code is not uniquely determinative.[17] No one has ever created an organism out of raw DNA. Even if this creation should come to pass, which may be conceivable for very "simple" organisms but much more remote for multicellular organisms, other parts of the cell participate in reproduction and development, and significantly modulate the role of the DNA in ways that are not DNA controlled.[18]

There is also abundant evidence that genes do not act in isolation but in systems.[19] It is not unreasonable to think of a cell as a "little ecosystem." Insertion of foreign genetic sequences does not merely add new function, nor does it leave the cell otherwise undisturbed. Instead, this genetic modification has the potential to create widespread alterations in gene expression patterns.[20] Mere knowledge of the genetic code does not even begin to give scientists the capacity to predict these types of systemic changes.[21] It is therefore the case that genetic modification has the potential to alter cellular metabolism in ways that we can neither understand, predict, nor control.[22] This unpredictability is not simply due to the complex interconnections within the genome. It is also due to the essentially random fashion in which genetic material is introduced. From this vantage point genetic engineering is not so much a technology as a gamble. And just as in Las Vegas, most of the wagers fail. Very few attempts to engineer organisms produce viable outcomes; the few that do survive often have significant problems emerge during the organism's life course or after reproduction.

The transfer of genes that are supposed to "code for" specific traits fails to account for the fact that genes and gene products are modified in ways that are specific not only to individual species, but also to particular tissue types within species.[23] Genes may play different roles when they are transferred into novel organisms than they play in the species from which they came. Thus, particularities of species and even tissues haunt and constrain genetic universalism. We can thus conclude that knowledge of the genetic code, while it provides new ways to manipulate organisms, does not go very far in helping us understand how organisms are affected by these manipulations.

This lack of knowledge, understanding, and control at the molecular and cellular level has ramifications when these techniques are applied to agricultural crops. Inserting a gene to add a desired characteristic—such as herbicide tolerance, frost tolerance, or salt tolerance—may have results other than the ones desired. First, the inserted genes may not function as intended, or may function optimally in only a narrow range of environmental conditions.[24] Beyond this, the organism may have unexpected additional metabolic alterations, some of which may lead to health risks such as allergenicity or toxicity in food products, or to detrimental effects on other organisms. These possibilities have finally been acknowledged even by the U.S. Food and Drug Administration,[25] after years of its insistence that genetically engineered foods were "substantially equivalent."[26] The significant likelihood of these complications contributes greatly to the enormous cost of developing viable genetically engineered varieties. This huge cost further belies public relations claims that genetic engineering of food is a practical, economical, people-oriented solution to world hunger.

Another difference from traditionally bred organisms has to do with gene silencing. The inserted genes may be modified or silenced by the organism. This can occur variably in different parts of the plant, and among different plants, and can worsen over the course of the growing season.[27] Such erratic gene expression deviates strikingly from that of traditionally bred organisms and their native genes. It indicates a potential serious intrinsic instability in genetically modified organisms. Such instability forebodes worrisome potential complications, particularly insofar as we allow our food supply to become dependent on these crops.

While some studies have demonstrated that these possibilities may occur, independent researchers are not generally funded to do these kinds of studies. Contrary to the complacent popular belief (in the United States at any rate) that our foods are well regulated, genetically engineered organisms are generally only tested by the companies that produce them, and these tests are reviewed fairly uncritically by regulators. We must ask whether we can entrust industry-sponsored or even industry-influenced science to seek evidence of such problems, let alone publicize such evidence if they find it.[28] Such results would be bad news for the bottom line; thus, between obedient/intimidated company employees and growing corporate influence on public research, the likelihood is minimized that such results, if obtained, will see the light of day.

The recourse to genetic modification of food crops to solve agricultural problems is yet another attempt to solve complex problems with a simple "magic bullet." Agriculture itself is a peculiar modification of growing patterns in the wild. In its currently dominant "industrial" forms, it tends toward monoculture, or at least toward a reduced number of coexisting organisms.[29] Many traditional agricultural systems, as well as contemporary organic and agroecological methods,[30] address not only the characteristics of individual species but also effects of intercropping on agricultural problems like pests and weeds. Industrial agriculture may attempt to fight infestations by applying or (in the case of GE food) inserting pesticides, but the efficacy is often at best modest, short lived, and rife with side effects such as toxicity and the emergence of resistance. In any event, organismic resourcefulness in getting around, adapting to, and defeating magic bullets is well established.[31] An agroecological approach to integrated pest management, on the other hand, which draws on intercropping and other interspecies interactions, can be safer, more effective, and more stable.[32] While some agricultural scientists see genetic modification as one tool in a larger agroecological armamentarium, the fact remains that genetic engineering techniques on their own are incapable of taking advantage of beneficial synergies in interspecies relationships. This is yet another way that interventions based in knowledge of gene code cannot in themselves provide a comprehensive basis for flexible agricultural practices. Thus, it is all the more disturbing that some advocates of genetic engineering in the developing world are a party to the dismantling of agricultural research stations that are not oriented toward genetic engineering.[33]

When one broadens the context even further to include ecology, the gene-based approach to crop organisms seems even more limited and short sighted. Genetic engineering is quite prone to creating ecological problems such as pollen flow to wild relatives, bioinvasion, and harm to other organisms through various direct and indirect pathways,[34] but it is incapable of solving or preventing these problems. Regarding biodiversity, the mode of operation of development of new genetically modified organisms tends to ignore rather than relate to local organism and ecology variants. Biotechnologists don't generally use scientific models that involve the interaction of organisms with specific ecological or cultural contexts. They tend to see biological features in a more general, context-independent way, rather than in relation to particular plants or animals that live in specific places with specific people. In addition, it is enormously expensive to produce GE food products, for one thing because it takes many thousands of laboratory failures before arriving at viable genetically modified varieties. There are thus multiple imperatives to market the seeds or animals that finally succeed in the lab in many widely differing ecological and cultural locales. Locally adapted varieties are displaced, in favor of GE varieties developed for ecologically unwise agricultural practices.

If we broaden the context still further and consider the diversity and cognitive richness of local cultures, we find that genetic engineering and industrial agriculture are blind to their integrity and value.[35] For industrial agriculture, the imperative of production predominates, and considerations such as the stabilizing and nurturing effects of relationships, community, and traditions have no meaning. These human and cultural structures appear as primitive obstacles to progress, which is defined by genetic engineers as a technologically facilitated bountiful harvest. But aside from the fact that genetic engineering's promise of improved yields is often not fulfilled,[36] there are further catastrophic impacts: farming communities are disrupted (particularly through the bankruptcy of smaller farms that cannot afford these technologies), and the accumulation of detailed local knowledge is lost.[37] Neither bounty nor genetic manipulation can substitute for what is destroyed. What remains is a homogenized and degraded countryside, cultural and material impoverishment, psychological devastation that passes from one generation to the next, and an abject dependency on multinational corporations.

THERE ARE OTHER POSSIBILITIES

Biotechnology, industrial agriculture, and genetic engineering of food are promoted as the only scientific options, but this is simply not true.[38] The science underlying these industrial approaches is actually primitive, outdated, and already surpassed.[39] Beyond the universalism of abstractions—here, GE's claim that there are no qualifications to the universality of the genetic code, and that context doesn't matter—there is a kind of science that is capable of incorporating what is known generally into an approach that is grounded locally.[40] Thus, opposition to genetic manipulation of food is not anti-science. Instead, the relentless press of genetic modification, shielded from critics, is retarding genuine scientific progress. This relentless press throws good money after bad in an attempt to recoup an investment that should not have been made in the first place, and marks a failure to be able to admit a mistake of this scale.

The right to food that is not genetically engineered is thus also a right to maintain allegiance to a different frame of reference from the productivist mentality and instrumentalist reductionism that genetic modification represents. The assertion of this right is much more than a meek demand for little preserves or reservations of organic farming in the midst of vast spreads of GE crops, or a tame request for GE-free labels on our food and GE-free aisles in our supermarkets. Certainly, demands for protecting organic farming and for food labeling have tactical importance. But they are not enough. (In any case, pollen spread appears to make it impossible to maintain crops that are organic and GE free in close proximity to the cultivation of genetically modified varieties.) The right to GE-free food is also important because the technology and its ideology are both immature and misguided. This makes it important that we not turn over the world's food supply to a poorly thought-out technological impulse. This right involves an insistence that there is a profound value to the panoply of particular, unique qualities of organisms, cultures, and ecosystems, and that they need to be protected from an inexorably destructive competitor. It also means we need to insist that this destructive competitor and interloper be stopped.

It does not appear, for the moment at least, that either the right to food free of genetic engineering or opposition to the right to pursue this questionable technology will be aggressively protected by governments or by international organizations such as the United Nations. These bodies are either in partnership with biotechnology corporations or have naively accepted biotechnology's claims that GE food is the best way to feed the world's hungry. Continuing grassroots pressure, an emerging awareness of agroecology as a scientifically informed rational approach that is more sophisticated than genetic engineering and industrial agriculture, and exposure of the hidden economic agendas of genetic engineering may help bring them around. GE crop failures or other disasters, should they occur and make it into the press, may contribute to this change of heart as well. Meanwhile the protection of the right to food free of genetic engineering remains for the most part a continual uphill battle against entrenched, unsympathetic, and incomprehending institutions.

The question remains whether in the long run genetic modification of food crops will find a more humble role in a truly ecologically and culturally friendly agricultural strategy. I would argue that the current technologies are intrinsically incapable of maturing in this fashion. While idealistic scientific agronomists may wish to incorporate genetic engineering into sustainable agriculture, they are unlikely to have grappled with the full range of objections to genetic engineering, and also are probably quite naive about the economic imperatives driving the biotech industry's commitment to this approach— imperatives that will hijack the goodwill of those who see positive applications of biotechnology. The prudent thing to do at this time, therefore, is to strengthen our opposition and to fight to preserve the knowledge bases and the biological and cultural diversity that we deeply need, given the unlikelihood that agricultural biotechnology will be transformed into a more modest, context-sensitive, and gentle technology.

The insistence on the right to food that is free of genetic engineering is both a plea and a struggle for human, organismic, cultural, and ecological viability. It rests on an understanding that the current generation of agricultural biotechnology was designed not with an appreciation of ecology and sustainability, but more with the aim of maximization of profits and production. Contemporary agricultural biotechnology unconstrained is gobbling everything in its path. Even worse, it may lead to major crop failures, because of the vulnerabilities arising from potentially unstable GE organisms applied as monocultures. Genetic engineering of food may thus create urgent needs for the very biological and cultural lifelines it is destroying. Protecting our right to food that is not genetically engineered will preserve some of these lifelines, and we may well need them.

NOTES

I would like to express my gratitude to the following people for their thoughtful and critical comments on various drafts of this chapter: Colin Gracey, George Scialabba, Chloe Silverman, Ruth Hubbard, Sheldon Krimsky, Peter Shorett, Abby Rockefeller, and Diana Cobbold.

1. Ivan Illich, *Tools for Conviviality* (New York: Harper, 1980).

2. Margaret Mellon, "The Wages of Hype: Agricultural Biotechnology After 25 Years," Arthur Miller Lecture presented at MIT (October 3, 2003); Marc Lappé and Britt Bailey, *Against the Grain* (Monroe, ME: Common Courage Press, 1998); Mae-Won Ho, *Genetic Engineering: Dream or Nightmare* (Bath, UK: Gateway Books, 1998).

3. Andrew Kimbrell, ed., *Fatal Harvest: The Tragedy of Industrial Agriculture* (Covelo, CA: Island Press, 2002).

4. M. S. Prakash and Gustavo Esteva, *Grassroots Post-modernism: Remaking the Soil of Culture* (London: Zed Books, 1998); Gustavo Esteva, "Re-embedding Food in Agriculture," *Culture & Agriculture*, Winter 1994, 2–12.

5. Frances Moore Lappe, Joseph Collins, and Peter Rossett, with Luis Esparza, *World Hunger: 12 Myths* (London: Earthscan, 1998); Miguel A. Altieri and Peter Rosset, "Ten

Reasons Why Biotechnology Will Not Ensure Food Security, Protect the Environment and Reduce Poverty in the Developing World," *AgBioForum* 2 (1999): 155–62, www.agroeco.org/doc/10reasonsbiotech1.pdf.

6. Richard Levins, "When Science Fails Us," www-trees.slu.se/newsl/32/32levin.htm (1996).

7. Barry Commoner, "Unraveling the DNA Myth: The Spurious Foundation of Genetic Engineering," *Harper's*, February 2002, 39–47.

8. Les Levidow and Susan Carr, "Unsound Science? Trans-Atlantic Regulatory Disputes over GM Crops," *International Journal of Biotechnology* 2 (2000): 257–73; B. Vogel and B. Tappeser, Der Einfluss der Sicherheitsforschung und Risikoabschätzung bei der Genehmigung von Inverkehrbringung und Sortenzulassung transgener Pflanzen, Öko-Institut e.V., study commissioned by the German Technology Assessment Bureau Auftrag, Berlin, 2000, available as a PDF file under www.oeko.de (only German). Also, see Jane Anne Morris, "Sheep in Wolf's Clothing," *By What Authority*, Fall 1998, www.poclad.org/bwa/fall98.htm.

9. Miguel Altieri, "Agroecology: The Science of Natural Resource Management for Poor Farmers in Marginal Environments," *Agriculture, Ecosystems and Environment* 93 (December 2002): 1–24, www.agroeco.org/doc/NRMfinal.pdf.

10. See USAID bilateral assistance programs, for example. Alan P. Larson, "The Future of Agricultural Biotechnology in World Trade," remarks at the Agricultural Outlook Forum 2002, www.state.gov/e/rls/rm/2002/8447.htm.

11. See Ian Sample, "Naïve, Narrow, and Biased," *Guardian*, Op-Ed, July 24, 2003; Sujatha Byravan, "Genetically Engineered Plants: Worth the Risk?" plenary lecture at Viterbo University, February 3, 2004.

12. See, for example, the "oncomouse" decision of the U.S. Patent and Trademark Office, U.S. Patent No. 4,736,866 (1988).

13. L. LaReesa Wolfenbarger and Paul R. Phifer, "The Ecological Risks and Benefits of Genetically Engineered Plants," *Science* 290 (2000): 2088–93.

14. Don Westfall, food industry marketing strategies consultant formerly with Promar International, quoted in Stuart Laidlaw, "Starlink Fallout Could Cost Billions," *Toronto Star*, January 9, 2001.

15. Emmy Simmons, assistant administrator, USAID, quoted in Philip Bereano, "Engineered Food Claims Are Hard to Swallow," *Seattle Times*, November 19, 2001.

16. See, for example, www.agroeco.org/doc/alt.maize-contam.pdf.

17. Evelyn Fox Keller, *The Century of the Gene* (Cambridge, MA: Harvard University Press, 2000).

18. Richard Lewontin, *Triple Helix: Gene, Organism, and Environment* (Cambridge, MA: Harvard University Press, 2000); Richard Lewontin, *It Ain't Necessarily So* (New York: New York Review of Books, 2000).

19. Commoner, "Unraveling the DNA Myth"; Ruth Hubbard and Elijah Wald, *Exploding the Gene Myth* (Boston: Beacon Press, 1993).

20. See Michael Hansen, "Genetic Engineering Is Not an Extension of Conventional Breeding," Consumers Union Discussion Paper (2000), www.consumersunion.org/food/widecpi200.htm. Also see David Schubert, "A Different Perspective on GM Food," *Nature Biotechnology* 20 (October 2002): 969.

21. Richard C. Strohman, "Organization Becomes Cause in the Matter," *Nature Biotechnology* 18 (June 2000): 575–6; Richard C. Strohman, "Five Stages of the Human Genome Project," *Nature Biotechnology* 17 (February 1999): 112.

22. Sui Huang, "The Practical Problems of Post-genomic Biology," *Nature Biotechnology* 18 (May 2000): 471–2.

23. See G. Riddihough and E. Pennisi, "The Evolution of Epigenetics," *Science* 293 (2001): 1063.

24. See, for example, P. Meyer, F. Linn, I. Heidmann, H. Meyer, I. Niedenhof, and H. Saedler, "Endogenous and Environmental Factors Influence 35S Promoter Methylation of a Maize A1 Gene Construct in Transgenic Petunia and Its Colour Phenotype," *Molecular Genes and Genetics* 231 (1992): 345–52.

25. Sheldon Krimsky, "Biotechnology at the Dinner Table: FDA Oversight of Transgenic Food," *Annals of the American Academy of Political and Social Science* 584 (November 2002): 80–96.

26. Erik Millsone, Eric Brunner, and Sue Mayer, "Beyond Substantial Equivalence," *Nature* 401 (October 7, 1999): 525–6.

27. Hansen, "Genetic Engineering Is Not an Extension of Conventional Breeding." Also see Meyer et al., "Endogenous and Environmental Factors Influence 35S Promoter Methylation"; and A. N. E. Birch, I. E. Geoghegan, D. W. Griffiths, and J. W. McNichol, "The Effect of Genetic Transformations for Pest Resistance on Foliar Solandine-Based Glycoalkaloids of Potato (*Solanum tuberosum*)," *Annals of Applied Biology* 140 (2002): 134–49.

28. Sheldon Krimsky, *Science in the Private Interest* (Lanham, MD: Rowman & Littlefield, 2003).

29. Kimbrell, *Fatal Harvest*; Wes Jackson and Wendell Berry, *New Roots for Agriculture* (Lincoln: University of Nebraska Press, 1985).

30. Miguel A. Altieri, *Agroecology: The Science of Sustainable Agriculture* (Boulder, CO: Westview Press, 1995).

31. See Pesticide Action Network North America, www.panna.org.

32. Altieri, *Agroecology*.

33. Fred Pearce, "Cashing in on Hunger: Biotechnology's Bid to Feed the World Is Leaving Less Profitable Techniques Starved for Funds," *New Scientist*, October 10, 1998.

34. Jane Rissler and Margaret Mellon, *The Ecological Risks of Genetically Engineered Crops* (Cambridge, MA: MIT Press, 1996).

35. Wendell Berry, *The Unsettling of America: Culture and Agriculture* (San Francisco: Sierra Club Books, 1977).

36. Mellon, "The Wages of Hype."

37. Stephen B. Brush and Doreen Stabinsky, eds., *Valuing Local Knowledge* (Covelo, CA: Island Press, 1996); Vandana Shiva, *Biopiracy* (Boston: South End Press, 1996).

38. Amory B. Lovins and L. Hunter Lovins, "A Tale of Two Botanies," *St. Louis Dispatch*, August 1, 1999, www.global-vision.org/misc/twobotanies.htm.

39. Lovins and Lovins, "A Tale of Two Botanies"; Commoner, "Unraveling the DNA Myth"; Martha Herbert, "Genetics Finding Its Place in Larger Living Schemes," *Critical Public Health* 12 (2002): 221–36.

40. Levins, "When Science Fails Us"; Steve Lerner, *Eco-Pioneers: Practional Visionaries Solving Today's Environmental Problems* (Cambridge, MA: MIT Press, 1997); Kenny Ausubel, *The Bioneers: Declarations of Interdependence* (South Burlington, VT: Chelsea Green, 2001); Alan Weisman, *Gaviotas: A Village to Reinvent the World* (South Burlington, VT: Chelsea Green, 1995).

• 7 •

A Right to GE-Free Food: The Case of Maize Contamination

Doreen Stabinsky

$\mathcal{I}$n attempting to address the right[1] to GE-free food, this chapter begins in a slightly indirect way—by telling a story about one of our most important food crops: corn, or as it is known to the rest of the world beyond the borders of the United States, maize.[2] But the chapter ends up demonstrating that for the Mexican peasants who grow maize, this right is not about whether they can walk into a grocery store and choose GE-free products; it is about whether they can continue to grow GE-free maize when 5 million tons a year of U.S. GE maize get dumped into their food and seed supply. In this case, the "right to GE-free food" means the collective right of agriculturalists to protect their traditional varieties and their *comida*[3] from transgenic contamination. As you read this chapter, keep in mind an underlying question that is examined in the concluding sections: what does a right to GE-free food really mean when there is no possibility of coexistence between GE and non-GE crops—when the mere acceptance of genetic engineering of the food supply spells the death knell for a GE-free existence?

THE CONTAMINATION OF MAIZE

In the fall of 2001, the Mexican government announced that traditional varieties of maize, growing in the far reaches of the mountainous state of Oaxaca, had been found to be contaminated with transgenic sequences. The findings were published in a controversial article in *Nature* magazine[4] and have been subsequently verified by Mexican government scientists.

This was no simple discovery—for years scientists had warned about the potential for contamination of valuable traditional varieties by transgenic sequences. Mexico is the most important center of diversity for maize in the world; traditional varieties are an important component of that diversity. In the early days, before solving world hunger was on their top-ten list of problems that genetic engineering would solve, biotech companies made promises that the technology would not be imported into centers of diversity so as to avoid the potential contamination that would ensue.

Why has this contamination been such a shock to the world community? Why should we care about contamination of farmer seed varieties in some of the poorest, most rural and isolated communities in Mexico? One important reason is the value to the world of the information contained in traditional crop varieties cultivated by indigenous farmers.

THE VALUE OF DIVERSITY

Cary Fowler and Pat Mooney won the Right Livelihood Award in 1985 for their work in attempting to halt the dangerous decline in crop diversity that was caused by the introduction into developing country agricultural systems of "improved" seeds from the developed world.[5] Apart from the problem that these "improved" seeds were usually not designed for the marginal environments in which peasant farmers in many countries operate, their promotion by agricultural ministries, development agencies, and transnational seed and chemical companies was threatening the existence of local varieties of crops. These local varieties contain a storehouse of genetic information useful not only to the peasants who developed them and continue to cultivate them, but to all who eat food on this planet. Plant breeders in the developed world continually look to traditional varieties to help them deal with constantly evolving weeds, insects, and diseases.

As Fowler and Mooney point out, these traditional varieties are our future. When they are lost, so eventually are we. In the 1970s and 1980s, they called attention to the loss of these varieties to homogenizing forces of developed country agriculture—dubbed genetic erosion. Now they call attention to the threats posed by invasion of transgenes into the traditional varieties.[6]

UNKNOWN IMPACTS OF CONTAMINATION

At present we have no data with which to determine the impacts of such contamination. The traditional scientific funding agencies in the United States—

such as the U.S. Department of Agriculture, National Science Foundation, and the like—have not adopted priorities to fund the research. Neither have the biotech companies—such as Monsanto, Syngenta, and Bayer—whose products contain the genes that have migrated into the Mexican varieties.

What we do know are the kinds of questions that should be asked in a scientific evaluation of impacts; we have knowledge of a range of unintended and unwanted effects that might be seen. Scientific study of several environmental impacts of the two main types of transgenic maize—*Bt* maize and herbicide-tolerant maize—demonstrate cause for concern in at least three areas: impacts on non-target organisms; impacts on soil fertility; and impacts on human health.

In a now famous article published in *Science* magazine, John Losey of Cornell University demonstrated that the pesticide engineered into *Bt* maize was of sufficient concentrations in maize pollen to kill that favorite American butterfly, the monarch.[7] Further field research has alleviated the concern somewhat—the field-level concentrations found in current commercial varieties do not appear to be toxic to monarchs over the short term, although no data exist to make statements about long-term impacts.[8] However, field-level concentrations of a variety removed from the market, *Bt*176, were toxic to monarchs.[9] What are the concentrations being expressed in the transgenic traditional maize varieties in Mexico? What are the non-target organisms in the mountain ecosystems of Oaxaca that are being exposed to the pesticide? What are the long-term effects of pesticide exposure on those non-target organisms? The answers to these questions are currently not known.

We can ask similar questions about organisms that live in the soil. Studies have shown that the *Bt* pesticide is exuded from maize roots and can exist for close to a year in some soil types.[10] We have no knowledge of impacts of the pesticide on the soil organisms living in Oaxacan soils. We do know that peasant farmers are absolutely dependent on the fertility of their soils, and by extension on the living organisms in the soil that create that fertility.

Perhaps the most troubling questions still unanswered about the maize contamination have to do with human health impacts. *Bt* is a pesticide produced by a soil microorganism and has long been used in organic agriculture. In this form it has been shown to be quite safe for farmers, consumers, and the environment. But genetic engineering takes the gene responsible for the pesticide, alters it, and expresses it in a foreign environment. The form of the protein found in the engineered plant is not the same as that found in the soil microorganism, yet toxicity testing has never been done on the protein as it is found in the new plant. In these mountain communities in Mexico, humans are not only consuming a novel protein; they are potentially consuming it in quantities far beyond what they have ever consumed before.

Consider the peasant agriculturalist diet in Mexico: large amounts of maize, processed only slightly, to make tortillas and other maize products that form the bulk of their diet. Toxicity studies in the United States have taken into consideration our own consumption patterns of genetically engineered corn, in highly processed goods such as taco shells and corn flakes. The safety of *Bt* maize consumption has not been evaluated based on the diet of Mexican—or African—peasants who consume large quantities of maize and who may already have a compromised health status due to malnutrition or parasites.

But *Bt* is only one of many proteins with which engineers plan to endow transgenic plants. Currently in commercial production in the United States are maize plants that produce a variety of human and animal drugs and industrial chemicals. For Norman Ellstrand of the University of California at Riverside, these pose the greatest risk to maize diversity and to humans. In a hypothetical scenario laid out by Professor Ellstrand, traditional varieties of maize become contaminated with a transgenic protein that has a chronically toxic effect on humans, which is discovered long after the gene has found its way through maize populations around the world. In his low-probability yet plausible scenario, the maize genome becomes irreversibly polluted and we cease being able as a species to consume maize.[11]

Unfortunately, human ignorance of organisms and ecosystems is great. This is a small sampling of effects that may arise over the next few generations; we are unlikely to be able to predict more than a fraction of the most significant consequences.

THE CULTURAL ROLE OF MAIZE

As much as we can understand, in a self-serving way, the importance of maize genetic diversity to the future of humankind, it is essential to get beyond a utilitarian vision of the genetic resource to understand the breadth of its significance to the multitude of peoples who consume maize. For the significance of the contamination is probably first and foremost a cultural affront of indescribable magnitude, a new colonial imposition of the biotechnological kind.

Maize is a cultural icon of the Mexican people. It is featured in their earliest stories about themselves, and in fact in the stories of many peoples of Latin America who have cultivated maize for millennia. One particularly beautiful creation myth, told in the *Popul vuh* of the Guatemalan Maya, tells the story of how the gods made man and woman out of maize.

Maize, according to Bonfil,[12] was more than domesticated—the maize plant was created by human effort. More than this, as the cultures of Mexican peoples evolved, it evolved along with their main foodstuffs. "At the same time they were cultivating maize, humans were cultivating themselves. The great civilizations of the past, the millions of Mexicans today, have as their root and foundation the generous maize. . . . Truly maize is the foundation of popular Mexican culture."[13]

In a recent manifesto to the NAFTA Commission for Environmental Cooperation, the signatories—most from the state of Oaxaca, where the contamination was first found—tried to convey the relationship that exists between maize and those who cultivate and eat it.

> We are people of maize. The grain is our brother, foundation of our culture, reality of our present. It is in the center of our daily life. It is a constant part of our diet and appears in one-quarter of the products that we buy in the stores. It is the heart of rural life and a constant ingredient in urban life. . . . Maize is more than a cereal. It sums up our past, defines our present and is the basis for a just future. We eat it, but it is not only food. It is a reason for fiestas, for exchange, for coexistence, for mutual help. It is our life.

The signatories to the manifesto reject transgenic maize—its imperial origin, its threat to their environment, its affront to their cultural heritage. "In our territories there will be no transgenics" is the concluding section of the statement.

A RIGHT TO GE-FREE FOOD?

I want to suggest that this is a more significant and profound question than it looks at first glance. When we ask the question from the comfort of our U.S. lifestyle and armchairs, it seems a pretty straightforward query: should consumers have the right to purchase GE-free products in their local supermarkets, the right to have labels on their food, the right to choose? But when we pose this question from the vantage point of the mountains of Oaxaca, we are asking about much more than labels. We, from the culture of McDonald's, are challenged to understand what it means to say "maize is my brother," and what transgenic contamination might mean to someone who would say that.

Regardless of our cultural origins, we can pose several questions here that highlight the problems inherent in maize contamination for both the people of maize and the people of McDonald's.

- What does it mean for all of us when there is no "away"? When pollen and seed travel freely and when it is clear to scientists that there is very little chance we can keep the maize genome uncontaminated?
- What does it mean when this contamination happens without our knowledge, without our assent, and against the wishes of the very

stewards of maize diversity—those peasants in Mexico who cultivate and reproduce their traditional varieties year after year?

- What does it mean for the culture of seed-saving and seed-sharing, when some communities in Mexico, with contaminated seed stocks, might decide to share their seeds no longer? What does this mean for spread and continuing enhancement of diversity among traditional varieties in Mexico?
- What does it mean for persons who consider themselves the sons and daughters of maize to suddenly have their totem food crop irreversibly contaminated with the latest fad of modern biology?
- How can it be that poor people on the margins of existence are once again serving as our guinea pigs for an unproven, risky, experimental technology?

There are many more questions like these waiting to be asked. Feel free to add to this list.

WE CAN'T GET THERE FROM HERE

Nothing in our experience to date shows that coexistence between GE and GE-free agriculture can work. Pollen moves. Plants are living, reproducing organisms. Before we even thought to look for contamination in Mexico, two different transgenes were spreading through the mountains of Oaxaca. Just this year, the tomato genetic conservation center of the University of California at Davis discovered that one of its varieties had been contaminated seven years prior and that they had dispersed the seeds of that variety to thirty-four researchers in fourteen countries around the world. Biotechnology and plant breeding companies (unfortunately the same entities) continuously struggle to keep seed lines pure; year after year, new tales of seed contamination are told.[14] New proposals for containment technologies are continually added to the first one—"terminator" technology—which prevents new grains from germinating and thus prevents farmers from saving seed. Unfortunately, even if pollen can be contained in the field, seed stocks can be accidentally contaminated, and environmental and human health effects will still remain.

Some would say this is a clear situation for applying the precautionary principle. The precautionary principle is an overarching principle for decision-making in the face of scientific uncertainty. Where there are potential threats to the environment or human health that may be severe or irreversible, decision-makers have an obligation to take action to prevent the harm that may ensue, even if there is not complete understanding of the nature, extent, or severity of those threats. Because engineered crops reproduce and disperse in the environment, the damage they can cause will be irreversible. Moreover, spread of damage with the spread of the engineered organism or transgene could result in an ecological catastrophe of unknown scope.

Even if we only want time to debate whether or not GE-free food is a "right," we need to put a halt to the genetic experiment right now. Five million tons of maize from the United States enter Mexico each year—the original source of the contamination of traditional varieties. At least 30 percent of that maize is genetically engineered. U.S. officials boast that so much GE maize is being cultivated in South Africa and shipped around the African continent that there will be no need to worry about GE-free areas in a few years—none will exist. The United States is assisting in the distribution of GE maize throughout the continent with its food aid programs. Monsanto and the U.S. government have forced open the Philippines as their beachhead in Asia to the commercial cultivation of GE maize.

We must call this situation what it is—a situation of plunder for profit, of corporations willfully polluting our food supply, polluting the food supply of peoples entirely dependent on what they grow, polluting the genetic resources upon which future generations will depend. The consequences of such pollution are mostly unknown. What we are sure of is that such pollution is irreversible. Regardless of whether we formulate the question in terms of protecting a cultural heritage or the right to buy non-GE foods in the supermarket, the answer is clear: stop the genetic experiment.

NOTES

1. While I do not here engage with the problematic concept of rights, I have to acknowledge the politically and culturally contestable nature of universal "rights," a Western philosophical construct seen as yet another method for colonizing cultural and political space. See the excellent discussion by Gustavo Esteva and Madhu Prakash, *Grassroots Postmodernism* (London, UK: Zed Books, 1998).

2. I use the words *corn* and *maize* interchangeably throughout the chapter.

3. *Comida* is the Spanish word for food. Its connotations are much broader than in English, however, and the meaning also embraces the sharing of food and of community that takes place during a meal.

4. David Quist and Ignacio Chapela, "Transgenic DNA Introgressed into Traditional Maize Landraces in Oaxaca, Mexico," *Nature* 414 (2001): 541–3.

5. See Cary Fowler and Pat Mooney, *Shattering: Food, Politics and the Loss of Genetic Diversity* (Tucson: University of Arizona Press, 1990).

6. ETC Group, "Fear-Reviewed Science: Contaminated Corn and Tainted Tortillas—Genetic Pollution in Mexico's Centre of Maize Diversity," January 23, 2002, accessed May 31, 2004, www.etcgroup.org.

7. J. E. Losey et al., "Transgenic Pollen Harms Monarch Larvae," *Nature* 399 (1999): 214.

8. Six separate papers on the topic were published in *Proceedings of the National Academy of Sciences* 98 (2001): 11908–942.

9. A. R. Zangerl et al., "Effects of Exposure to Event 176 *Bacillus thuringiensis* Corn Pollen on Monarch and Black Swallowtail Caterpillars Under Field Conditions," *Proceedings of the National Academy of Sciences* 98 (2001): 11908–912.

10. D. Saxena and G. Stotzky, "Insecticidal Toxin from *Bacillus thuringiensis* Is Released from Roots of Transgenic *Bt* Corn *in vitro* and *in situ*," *FEMS Microbiology and Ecology* 33 (2000): 35–9.

11. Norman Ellstrand, "After Centuries of Introgression from Cultivated Plants to Wild Relatives, What's Next?" keynote presentation at the conference "Introgression from Genetically Modified Plants (GMP) into Wild Relatives and Its Consequences," sponsored by the European Science Foundation AIGM Program, January 21–24, 2003, Amsterdam, Netherlands. In his presentation, Ellstrand calls this scenario "very unlikely but not impossible."

12. As cited by G. Esteva, "Introducción," in G. Esteva and C. Marielle, eds., *Sin maís no hay país* (Mexico City: Culturas Populares de Mexico), 11.

13. As cited by G. Esteva, "Introducción." Translated from original by author.

14. In 2000, Friends of the Earth International found that an unapproved GE maize variety—StarLink—had contaminated a range of food products in the United States. Four years later, the U.S. government testing program still finds traces of StarLink in the seed and food supply. In April 2001, controls by the Ministry of Environment of Schleswig-Holstein, Germany, found GE contamination in Monsanto's maize variety Arsenal. Also in 2001, testing commissioned by Greenpeace Austria found GE contamination in Pioneer maize seeds. Numerous other scandals of contamination in cotton, maize, and canola seed have been documented around the world.

Ensuring the Public's Right to Safe Food

Richard Caplan

$\mathcal{A}$ccording to the most relevant dictionary definition, a right is "something to which one has a just claim." Proponents of genetically engineered crops would say that if genetically engineered crops are as safe as their conventional counterparts—if we knew that the environmental and human health risks were negligible—then it would not be fair to say that we have a right to a food supply free of genetically engineered material. In that scenario, labeling and segregation would be unnecessary, as the costs would not be justified to separate goods indistinguishable in terms of their impact on the environment and human health. This chapter explores whether we have a right to a food supply free of genetically engineered material by examining the claims of proponents and critics of the technology.

In fact, the important differences between conventional crops and genetically engineered crops merit a food supply that can be guaranteed free of genetically engineered material. But even if demonstrations of risk from the technology had not been clearly identified, the right to a food supply free from genetically engineered material must still exist. Genetic pollution is irreversible, and given the poor track record of the companies pushing genetically engineered crops without regard to human health and environmental safety, citizens deserve to know that safety is an option in the future.

The foremost argument of proponents of genetically engineered crops against a right to a food supply free of genetically engineered material is that genetically engineered crops are well tested and have been demonstrated to be safe for human health. A generally accepted standard for establishing a fact through research is to perform studies and subject them to the rigors of peer review. Biotechnology companies have quite dramatically failed to meet this standard, and the policies of the U.S. Food and Drug Administration (FDA) have allowed them to do so. Not only does the FDA not require any mandatory safety testing for genetically engineered crops, but the agency does not even require mandatory notification by companies seeking to commercialize them.[1] When a tepid proposal was published in the *Federal Register* to require mandatory notification but not mandatory safety testing,[2] it was eventually withdrawn,[3] and the system in place remains entirely voluntary.

Companies generally do not send complete studies they perform to regulatory agencies in the United States, only summaries. Nor do they publish studies in mainstream scientific journals. When *Science* magazine recently published the results of a literature search conducted for all peer-reviewed articles that assessed the safety of genetically engineered foods, a mere handful turned up.[4] That number was dwarfed by the number of articles proclaiming the safety of the technology based on mere assertion, not fact. Given the dearth of empirical evidence in the scientific literature on the safety of genetically engineered crops, any claims of safety are premature.

Another claim in support of genetic engineering technology is that it will usher in an era of ecologically sustainable agriculture, reducing or eliminating toxic farm chemicals, reducing topsoil erosion, and producing more food on less land. While these promises appear noble, they are thus far largely empty. By far the most widely planted genetically engineered trait is herbicide tolerance, wherein a crop is engineered to withstand blasts of herbicide designed to kill all other flora but the plant. This technology has resulted in an increased use of herbicide, not a decrease.[5] And even as the chemical argument is fought out, with proponents of genetic engineering claiming chemical reductions in the short term from *Bt* cotton, and certainly the possibility of future products that could reduce chemical use, the reality is that genetic engineering is perpetuating an agricultural paradigm antithetical to true ecological agriculture. The emphasis in genetic engineering is on large-scale monoculture agriculture, either depending on herbicides, or replacing the failed "one-pesticide-one-pest" model with a "one-gene-one-pest" model. A generation ago, initial support for pesticides came from farmers, the government, and certainly the chemical companies. All have been forced to retreat from their early euphoria. Now that the same companies that brought us the first generation of toxic farm chemicals have reinvented themselves as biotechnology companies, it is with justifiable skepticism that we view their latest claims of environmental stewardship.

The biotechnology industry has focused its costly public relations campaign on dubious claims that this technology has a fundamental role to play in feeding the world. The claim that a genetic quick fix from a laboratory could solve the world hunger problem was debunked many years ago. For example, in his book *Selfish Gene*, Richard Dawkins writes: "Increases in food production may temporarily alleviate the problem, but it is mathematically certain that they cannot be a long-term solution; indeed, like the medical advances that have precipitated the crisis, they may well make the problem worse, by speeding up the rate of the population expansion." Peter Rosset of Food First has pointed out that we have enough food already available to provide 4.3 pounds to every person every day; the real problems are poverty and inequality. And Tewolde Berhan Gebre Egziabher of the Institute of Sustainable Development in Addis Ababa, Ethiopia, has stated the problem perhaps better than anyone. "There are still hungry people in Ethiopia, but they are hungry because they have no money, no longer because there is no food to buy. We strongly resent the abuse of our poverty to sway the interests of the European public."[6] While both conventional breeding and genetic engineering can increase yields, neither approach will necessarily have anything to do with addressing the more fundamental, root causes of hunger. Far too many people still suffer from hunger in the United States alone, despite agricultural surpluses, and thus it is naive to argue that merely increasing food production will necessarily alleviate hunger.

Finally, those advocating genetically engineered crops have begun openly supporting a more nefarious strategy of foisting their products on an unwilling public and intentionally contaminating the food supply with genetically engineered material. For example, Dave Adolphe of the Canadian Seed Growers Association was quoted in the *Western Producer* in April 2002 stating, "It's a hell of a thing to say that the way we win is don't give the consumer a choice, but that might be it." This argument is tied to the claims that consumer choice is neither technically nor economically feasible. Proponents of the technology, without putting forward facts, are claiming consumer choice is not possible, while working diligently to make it impossible through a campaign of deliberate pollution. This strategy is utterly reprehensible.

Thus far, this chapter has addressed the arguments of proponents of genetically engineered crops and exposed them as unconvincing. A review of the major criticisms of the technology will further clarify the need for a right to a food supply free of genetically engineered material. The analysis will cover the lack of long-term testing, the lack of adequate regulatory control, and the right to opt out of a system of corporate agriculture.

The scientific literature reveals that genetically engineered crops have not been well tested, nor tested in any way that can be described as long term, for their impact on human health or the environment. As mentioned above, a literature review of human health studies turned up far too few references to allow for any legitimate conclusions. Given that many processed foods in American supermarkets already contain genetically engineered material, usually in the form of processed soy or corn, this is a serious problem, as the range and severity of risk posed by the technology is noteworthy, including allergenicity, toxicity, nutritional changes, and exacerbation of antibiotic resistance.[7] At a time when these issues are not resolved, eating labeled genetically engineered crops is premature. Instead, these products should not be on the market. A similar review in *Science* for research into environmental impacts found many studies but could only conclude that none of the benefits touted by proponents of genetically engineered crops could be confirmed, and that none of the risks had been disproven. While some of the risks to the environment can be studied and resolved, there is a degree of inherent unpredictability to the technology that has led some of the country's most respected scientists to call its current intellectual foundation "spurious."[8]

Because of the risks of genetically engineered crops, the public should expect a proactive governmental response and the creation of a regulatory system that protects human health and the environment. The system for oversight of genetically engineered crops in the United States is emphatically not that system. With respect to regulations governing genetically engineered crops, the Food and Drug Administration operates under a vague statement of policy published in 1992. The FDA regulations were developed out of statutes conceived and written well before the technology was available. The FDA, for example, even admits that it "does not conduct a comprehensive scientific review of data generated by the developer [of a genetically engineered food]."[9] The Environmental Protection Agency allowed the commercial planting of millions of acres of *Bt* crops before their regulations were ever finalized. And the Department of Agriculture (USDA) recently had their oversight criticized by the National Academy of Sciences, which found that the management of GM crops by the USDA was at times "scientifically inadequate" and "superficial," and that their reviews "lacked scientific rigor, balance, and transparency."[10] Oversight of genetically engineered crops in the United States is in some ways severely insufficient, and in others almost lacking entirely.

Genetically engineered crops that have been commercialized are part of the perpetuation of a failed agricultural paradigm of large-scale monoculture and intensive chemical inputs. The overwhelming majority of genetically engineered crops are engineered to withstand blasts of herbicide, a trait that has resulted in an increase in the use of herbicides. Regardless of the impact on chemical use in the short term, it is clear that the technology is being used to further the movement toward just a few elite strains of each of the major crops (corn, soy, cotton, etc.), a shift that erodes genetic diversity and leaves crops vulnerable to attack (both natural, as in the 1970 Southern Leaf Blight episode, or anthropogenic, from a malicious biological attack). Many consumers are making it increasingly clear that they want a different kind of food system, as evidenced by, for example, the growth of local food policy councils, farmer's markets, community supported agriculture systems, and organic agriculture.

There may be genetically engineered crops in the future, when enough evidence accumulates that such crops can be said to present no unreasonable risk to human health or the environment. That is not the case for the crops that have currently been commercialized. For any crops that do meet the threshold of safety for human health and the environment, consumer interest and consumer demand clearly indicate a need for a system of choice that allows people to know, to the best of our ability, where their food was grown, who grew it, what they grew, when they grew it, why they grew it, and how they grew it. For genetically engineered crops, the uncertain science supporting the technology demands that consumers have a right to these answers. The risks surrounding the technology demand that this be a right. Through their purchasing decisions and in opinion poll after opinion poll, consumers have made it clear that they want this right. It can and should be granted.

NOTES

1. Richard Caplan, "Failure to Do Anything: Regulation of Genetically Engineered Foods at FDA" (October 2000). Available at pirg.org/ge/GE.asp?id2=4781&id3=ge&.

2. U.S. Department of Health and Human Services and Food and Drug Administration, "Premarket Notice Concerning Bioengineered Foods," *Federal Register* 66:12 (January 18, 2001).

3. U.S. Department of Health and Human Services, "Semiannual Regulatory Agenda," *Federal Register* 68:101 (May 27, 2003).

4. Jose L. Domingo, "Health Risks of GM Foods: Many Opinions but Few Data," *Science*, June 9, 2000.

5. Charles Benbrook, "Troubled Times amid Commercial Success for Roundup Ready Soybeans," AgBioTech InfoNet Technical Paper (May 3, 2001). Available at biotech-info.net/troubledtimes.html.

6. Totnes Genetix Group fact sheet. Available at www.togg.org.uk/resources/feedtheworld.html.

7. See, for example, Michael Hansen, "Science-Based Approaches to Assessing Allergenicity of New Proteins in Genetically Engineered Foods," presentation to FDA Food Biotechnology Subcommittee, Food Advisory Committee (August 14, 2002). Consumers Union's comments on Docket No. 00N-1396, "Premarket Notice Concerning Bioengineered Foods" (May 1, 2001). Available at www.consumersunion.org/food/biocpi501.htm.

8. Barry Commoner, "Unraveling the DNA Myth: The Spurious Foundation of Genetic Engineering," *Harper's*, February 2002.

9. U.S. Food and Drug Administration, *Guidance on Consultation Procedures: Foods Derived from New Plant Varieties* (Washington, DC: Center for Food Safety and Applied Nutrition, October 1997). Available at vm.cfsan.fda.gov/~1rd/consulpr.html.

10. National Research Council, *Environmental Effects of Transgenic Plants: The Scope and Adequacy of Regulation* (Washington, DC: National Research Council, 2002).

Part IV

INDIGENOUS PEOPLES

"All indigenous peoples have the right to manage their own biological resources, to preserve their traditional knowledge, and to protect these from expropriation and biopiracy by scientific, corporate or government interests."

Article 4 of the Genetic Bill of Rights

· 9 ·

Acts of Self-Determination and Self-Defense: Indigenous Peoples' Responses to Biocolonialism

Debra Harry

$\mathcal{I}$ndigenous peoples are heavily impacted by the ever increasing global interest in the gene business because we control and occupy lands that are rich in biodiversity.[1] Additionally, we maintain traditional knowledge about the beneficial uses of the flora and fauna in our territories, making this knowledge particularly useful to gene hunters seeking potentially profitable genetic resources.

Indigenous peoples are also subjects of a wide array of human genetic research projects because of the perceived uniqueness of our human gene pools. Over the past decade, we have experienced extensive violations of human rights by researchers who have failed to obtain informed consent before taking samples, or who have allowed widespread unauthorized secondary uses or commercialization of our genetic samples. For instance, in March 2004, the Havasupai Tribe of Arizona filed a $50 million lawsuit in Coconino County Superior Court against Arizona State University, the Arizona Board of Regents, and three ASU professors.[2] The suit claims that more than 400 blood samples were taken from tribal members between 1990 and 1994 putatively for diabetes research. Instead, the samples were used in research on inbreeding, schizophrenia, and theories about ancient human population migrations to North America. In 2002, members of the Nuu-Chah-Nulth Tribe in British Columbia were outraged to find that samples taken for arthritis research over twenty years ago were still being used in England for human population research without their consent.[3] It is important to recognize that the majority of health problems experienced by Indigenous peoples are caused by social/environmental factors and are not genetic.

Through the application of Western intellectual property law (patents and copyright), corporations are claiming ownership over genes, products, and data derived from genetic resources. The commercialization of genes conflicts with Indigenous values and the collective nature of our customary management systems. Indigenous peoples have been consistent in our calls for no patents on life forms, as expressed in this declaration issued in 1995 at a meeting of Indigenous groups from the Western Hemisphere:

> To negate the complexity of any life form by isolating and reducing it to its minute parts, western science and technologies diminishes its identity as a precious and unique life form, and alters its relationship to the natural order. Genetic technologies which manipulate and change the fundamental core and identity of any life form is an absolute violation of these principles, and creates the potential for unpredictable and therefore dangerous consequences. We oppose the patenting of all natural genetic materials. We hold that life cannot be bought, owned, sold, discovered or patented, even in its smallest form.[4]

Monopoly control over genes and germplasm threatens our food security and undermines the livelihoods and self-sufficiency of millions of people who depend on the seed saved from traditional agricultural crops for future harvest. Patents over agricultural resources have the potential to make seeds unaffordable, and genetically modified seeds raise additional concerns about the possibilities of genetic contamination of local crops or the environment.

This chapter examines the basis for the efforts by Indigenous peoples to assert our right of self-determination and to seek recognition and protection of our rights within the international human rights framework. In exercising the right of self-determination, it follows that Indigenous peoples have the right to own, manage, and protect our natural resources, including genetic resources, within our territories. Finally, this chapter discusses the concerns of Indigenous peoples regarding the application of intellectual property rights over genetic resources and traditional knowledge.

RELATIONSHIP OF GENETIC RESEARCH TO THE RIGHT OF SELF-DETERMINATION

Articles 1 (1) of both the International Covenant on Civil and Political Rights[5] and the International Covenant on Economic, Social and Cultural Rights[6] state: "All peoples have the right of self-determination. By virtue of that right they freely determine their political status and freely pursue their economic, social and cultural development." At the same time, it is widely recognized that states often deny or diminish the ability of Indigenous peoples to exercise the right of self-determination. They are threatened by Indigenous peoples' prior and paramount rights to our territories, which conflict with their subsequent interests over those same territories and resources.

In the report entitled "Indigenous Peoples' Permanent Sovereignty over Natural Resources," Special Rapporteur Erica Irene A. Daes notes: "in legal principle, there is no objection to using the term sovereignty in reference to Indigenous peoples acting in their governmental capacity, although that governmental capacity might be limited in various ways."[7]

The report further states:

> Though indigenous peoples' permanent sovereignty over natural resources has not been explicitly recognized in international legal instruments, this right may now be said to exist. That is, the Special Rapporteur concludes that the right exists in international law by reason of the positive recognition of a broad range of human rights held by indigenous people, most notably the right to own property, the right to ownership of the lands they historically or traditionally use and occupy, the rights to self-determination and autonomy, the right to development, the right to be free from discrimination, and a host of other human rights.[8]

The right of self-determination is the fundamental premise upon which Indigenous peoples have asserted our proprietary, inherent, and inalienable rights over our traditional knowledge and biological resources. We continue to maintain and protect biologically diverse ecosystems for the collective good.

Several international human rights instruments recognize the collective nature of Indigenous peoples' rights of self-determination, including the United Nations Draft Declaration on the Rights of Indigenous Peoples,[9] Convention 169 of the International Labor Organization,[10] and the Draft American Declaration on the Rights of Indigenous Peoples of the Organization of American States.[11]

By far, the UN Draft Declaration on the Rights of Indigenous Peoples is the international instrument that is the most representative of Indigenous thought and participation,[12] and constitutes "the minimum standards for the survival, dignity and well-being of the Indigenous peoples of the world." The Declaration states: "Indigenous peoples have the right to own, develop, control and use the lands and territories . . . which they have traditionally owned or otherwise occupied or used."[13] The Declaration further states:

> Indigenous peoples are entitled to the recognition of the full ownership, control and protection of their cultural and intellectual property. They have the right to special measures to control, develop and protect their sciences, technologies and cultural manifestations, including human and other genetic resources, seeds, medicines, knowledge of the properties of

fauna and flora, oral traditions, literatures, designs and visual and perform-
ing arts.[14]

INDIGENOUS PEOPLES' CALL TO
PROTECT GENETIC RESOURCES

At the international level, Indigenous peoples have continually asserted our inherent right to protect our cultural and natural resources. As the economic potential of genetic resources began to be realized in the early 1990s, genes were
a key topic discussed at the UN Conference on Environment in Rio de Janeiro
in June 1992, commonly referred to as the Earth Summit. The Earth Summit
established the Convention on Biological Diversity (CBD), and set into mo-
tion the international cooperation of states to ensure the "conservation of
biological diversity, the sustainable use of its components and the fair and equitable sharing of benefits arising from its utilization."[15]

Indigenous peoples also issued the Indigenous Peoples' Earth Charter,
known as the Kari-Oca Declaration, at the Earth Summit:

> We, the Indigenous peoples, maintain our inherent rights to self-determination.
> We have always had the right to decide our own forms of government, to use
> our own laws, to raise and educate our children, to our own cultural identity
> without interference. . . . We maintain our inalienable rights to our lands and
> territories, to all our resources—above and below—and to our waters. We as-
> sert our ongoing responsibility to pass these onto the future generations.[16]

Indigenous peoples have participated in the CBD and other UN
processes, consistently reiterating our fundamental demands for respect and
protection of our rights as peoples. This presence helps to hold the Parties accountable to the founding principles, even though it has become evident that
the proposals for the access and benefit-sharing of genetic resources have become the paramount objective of the Parties. In particular, the current efforts
to elaborate an international regime on access and benefit-sharing being discussed at the international level are mechanisms that will facilitate the exploitation of genetic resources.

The recent benefit-sharing agreement reached by South Africa's Council
for Scientific and Industrial Research (CSIR) and the San community is
widely hailed as a landmark agreement with the potential of earning the San
millions of rand for the use of their knowledge in the commercialization of the
hoodia plant. CSIR sold the development rights of the active ingredient, called
"P57," of the appetite-suppressing hoodia plant to Phytopharm, who in turn
licensed the right to Pfizer to develop the drug.

Rachel Wynberg of Biowatch SA, however, notes the agreement is not without problems: "The San will receive only a fraction of a percent—less than 0.003 percent—of net sales. The San's money will come from CSIR's share while the profits received by Phytopharm and Pfizer will remain unchanged." She further notes: "the agreement explicitly prevents the San from using their knowledge of Hoodia in any other commercial application."[17] This agreement was reached only after CSIR and Phytopharm were widely criticized for failing to get the consent of the San or recognize the role that the San's knowledge played in identifying hoodia's ethnobotanical properties.

Wynberg raises a moral dilemma for Indigenous peoples that is implicit in any benefit-sharing agreements based on the commercialization of traditional knowledge and patenting genetic resources. "In communities such as the San, the sharing of knowledge is a culture and basic to their way of life. The patenting of active compounds of Hoodia by the CSIR runs counter to this belief, yet brings with it greater financial returns—and greater risk—than the commercialization of non-patented herbal medicines." The San experience indicates that benefit-sharing agreements can coerce Indigenous communities to participate in the patenting of genetic resources, alienating them from the use of traditional knowledge for other benefits in a manner that conflicts with traditional values.

Thus, many Indigenous peoples fear that the global discussions on sustainable development and poverty alleviation are now replaced by discussions about new rules and mechanisms for the abject exploitation of genetic resources, all in the name of sustainable development. Recognizing that our traditional knowledge constitutes the collective heritage and patrimony of our peoples, we are refusing to place economic value on these things for exploitation.

The Beijing Declaration of Indigenous Women, issued at the UN Fourth World Conference on Women in Beijing, asserts the following demands:

> We demand that our inalienable rights to our intellectual and cultural heritage be recognized and respected. We will continue to freely use our biodiversity for meeting our local needs, while ensuring that the biodiversity base of our local economies will not be eroded. We will revitalize and rejuvenate our biological and cultural heritage and continue to be the guardians and custodians of our knowledge and biodiversity.[18]

In 2003, ten years after Rio, at the World Summit on Sustainable Development in Johannesburg, Indigenous peoples issued the Kimberly Declaration, stating:

> As peoples, we reaffirm our rights to self-determination and to own, control and manage our ancestral lands and territories, waters and other resources. Our lands and territories are at the core of our existence—we are the land and the land is us; we have a distinct spiritual and material

relationship with our lands and territories and they are inextricably linked to our survival and to the preservation and further development of our knowledge systems and cultures, conservation and sustainable use of biodiversity and ecosystem management.

Indigenous peoples have been adamant in our rejection of population-based genetic research projects. In the early 1990s initiatives such as the proposed Human Genome Diversity Project specifically targeted Indigenous populations for the collection of blood samples. Indigenous peoples were outraged at attempts by U.S. federal agencies to patent the human genetic material of Indigenous peoples from Panama and the Solomon Islands. When the U.S. Department of Health and Human Services was actually granted a patent over the cell line of a Hagahai man from Papua New Guinea[19] in 1994, it drew sharp international criticism. In a 1995 Declaration of Indigenous Peoples of the Western Hemisphere, Indigenous peoples called for "an immediate moratorium on collections and/or patenting of genetic materials from Indigenous persons and communities by any scientific project, health organization, governments, independent agencies, or individual researchers" and expressed "solidarity to all those who are . . . seeking the repatriation of genetic materials already taken."[20]

The Palapala Kulike OKa`Aha Pono Paoakalani Declaration, issued by Kanaka Maoli, the Indigenous peoples of the Hawaiian Archipelago, in October 2003 states: "Kanaka Maoli human genetic material is sacred and inalienable. Therefore, we support a moratorium on patenting, licensing, sale or transfer of our human genetic material."[21]

These declarations, developed by leaders from Indigenous nations around the world, represent the aspirations, advocacy, and expressions of the right of self-determination of Indigenous peoples.

IMPOSITION OF INTELLECTUAL PROPERTY RIGHTS ON GENETIC RESOURCE RIGHTS

The short-term and individuated nature of intellectual property rights conflicts with the collective rights, long-term protection, and customary management systems of Indigenous peoples. One author describes the conflict succinctly:

In particular, there is a very serious question whether the category "property," or the historically contingent and individualistic notion of property that has arisen in the West, is even appropriate when discussing things like agricultural practices, cell lines, seed plasm, and oral narratives that belong to communities rather than individuals. If we are not capable of acknowledging the existence of different life-worlds and ways of envisioning human beings' relationship to the natural world in our intellectual property laws, then unfortunately, it may be late in the day for biodiversity and hopes for a genuinely multicultural world.[22]

The Mataatua Declaration resulting from the International Conference on the Cultural and Intellectual Property Rights of Indigenous Peoples in June 1993 articulates the need for international standards of protection for Indigenous peoples' knowledge and resources. It encourages states to develop, in co-operation with Indigenous peoples, new protection mechanisms that reflect the following principles: "collective (as well as individual) ownership and origin, retroactive coverage of historical as well as contemporary works, protection against debasement of culturally significant items, co-operative rather than competitive framework, first beneficiaries to be the direct descendants of the traditional guardians of that knowledge, and multi-generational coverage span." And it calls upon the UN to "monitor and take action against any States whose persistent policies and activities damage the cultural and intellectual property rights of indigenous peoples."[23]

A problematic provision of the CBD for Indigenous peoples is contained in Article 15.1, which "recognizes the sovereign authority of States over their natural resources, and the authority to determine access to genetic resources rests with the national governments."[24] However, Article 8 (j) contains provisions requiring Parties to "respect, preserve and maintain knowledge, innovations and practices of indigenous and local communities relevant to conservation and sustainable use of biological diversity, and to encourage the fair and equitable sharing of the benefits arising from the utilization of knowledge, innovations and practices of indigenous and local communities."[25] Article 8 (j) has enabled Indigenous peoples to be active in subsequent deliberations regarding implementation of the CBD.

Indigenous peoples have also noted the link between the global hunt for genetic resources and international trade agreements. The Beijing Declaration of Indigenous Women states:

> We demand that the western concept and practice of intellectual property rights as defined by the TRIPS in GATT, not be applied to Indigenous peoples communities and territories. We demand that the World Trade Organization recognize our intellectual and cultural rights and not allow the domain of private intellectual rights and corporate monopolies to violate these.[26]

A Declaration of Indigenous Peoples of the Western Hemisphere issued in 1995 states: "We denounce all instruments of economic apparatus such as

NAFTA, GATT and the World Trade Organization (WTO) which continue to exploit people and natural resources to profit powerful corporations, assisted by governments and military forces of developed countries."[27]

The World Intellectual Property Organization[28] (WIPO) Intergovernmental Committee on Intellectual Property and Genetic Resources, Traditional Knowledge and Folklore suggests that current or modified forms of intellectual property rights (IPRs) can be used to protect traditional knowledge. Many Indigenous peoples, however, have been critical of the imposition of IPR regimes over our collective resources and knowledge.

A report on the "Workshop on Biodiversity, Traditional Knowledge and Rights of Indigenous Peoples" critically analyzes IPRs from an Indigenous perspective. The report notes that IPRs fail to acknowledge the customary systems of Indigenous peoples to safeguard and protect our knowledge and concludes that "intellectual property rights cannot and will not adequately protect traditional knowledge."[29]

One proposed strategy suggests that Indigenous peoples should document their knowledge in registries or databases in order to establish proof of prior art. In Indigenous territories, the primary means of protection and transmission of biodiversity-related traditional knowledge continues to be through customary laws, traditional practices, and oral histories. Traditional knowledge is dynamic, not static, and cannot simply be documented and "fixed in a tangible form" to meet intellectual property law requirements. When Indigenous peoples utilize databases to document their knowledge, they do this as a cultural preservation strategy and not for the purposes of establishing prior art.

Some have argued that

> Published data, or data previously taken from Indigenous communities, with or without consent, should be considered to be in the public domain and freely accessible for use by anyone. Indigenous peoples have asserted that "with respect to traditional knowledge that is already documented or in registers or databases, this knowledge should not be considered to be in the public domain, and Indigenous peoples retain all rights over the ownership and use of this knowledge. Similarly, any traditional knowledge acquired without prior and informed consent is not in the public domain, and all rights remain with the affected Indigenous peoples."[30]

Perhaps the most important distinction that separates Indigenous peoples from other ethnic and cultural groups is the fact that we have territorial rights. Indigenous peoples are "owners" of the resources that exist within their territories. It is "real" property in the eyes of Western property law, but more importantly it is recognized as a national resource and right by Indigenous peoples themselves. Without territory, Indigenous peoples know their cultural, political, and spiritual existence and livelihoods are threatened.

These beliefs conflict with the assertions by some progressive NGOs[31] that the world's genetic resources should belong to the commons. While many Indigenous peoples agree there should be no patents over genetic resources, the notion that the commons should also extend over our territories and governance systems has been rejected. As a result, these initiatives have failed to receive widespread endorsement by Indigenous peoples. Despite being the antithesis of each other, the commons and IPRs are both grounded in notions of Western property law. Asserting the commons over genetic resources is a useful goal, but its extension over Indigenous peoples' territories would constitute yet another act of colonization on Indigenous peoples. It nevertheless is a worthy effort to encourage states to restrict patents on genetic resources.

CONCLUSION

International law recognizes the right of "self-determination of *all peoples*" and does not contain an exclusionary note with reference to Indigenous peoples. International law is built upon a preeminent respect for the right of self-determination as the basis for international relationships. The principle of self-determination protects the rights of less powerful or newly emerging states from unfair exploitation of their natural resources by more powerful states.[32]

Indigenous peoples have never abandoned our right of self-determination. We are peoples, with our own collective histories, languages, cultures, social systems based upon our inalienable connection to our territories. As peoples, we are entitled to the same recognition and respect afforded others in the international community. Indigenous peoples are seeking international standards and mechanisms that ensure equity, justice, and respect for our collective rights. Until such standards are in place, Indigenous peoples will have to continue to exercise their self-determination to protect their knowledge and resources by any means possible at the local and national levels. Some have declared their territories life-form patent free zones,[33] while others have enacted local and national legislation to regulate research.[34] Others have issued statements asserting their proprietary rights over all resources in their territories.[35] Regardless of how the international debates are resolved, and given the lack of regulation prohibiting exploitive practices, Indigenous peoples must continue to exercise our right of self-determination by instituting strategies for the defense of our knowledge and resources.

NOTES

1. When discussing Indigenous peoples' rights, positions, or beliefs in this chapter, I write in the first person to reflect my identity as a Kuyuidicutta (Northern Paiute) from Pyramid Lake, Nevada.

2. "Havasupai Tribe Files a $50M Suit Against ASU," *Arizona Daily Sun*, March 16, 2004.

3. "Blood Promise," *CBC News Online*, September 27, 2000, vancouver.cbc.ca/cgi-bin/templates/view.cgi?/news/2000/09/27/bc_blood000927.

4. Declaration of Indigenous Peoples of the Western Hemisphere Regarding the Human Genome Diversity Project, Phoenix, Arizona, 1995, www.ipcb.org/resolutions/htmls/dec_phx.html.

5. United Nations, *International Covenant on Civil and Political Rights* (adopted December 19, 1966, entered into force March 23, 1976, 999 U.N.T.S. 171).

6. United Nations, *International Covenant on Economic, Social and Cultural Rights* (adopted December 19, 1966, entered into force January 3, 1976, 999 U.N.T.S. 3).

7. Erica Irene A. Daes, Final Report of the Special Rapporteur, "Indigenous Peoples' Permanent Sovereignty over Natural Resources," UN Economic and Social Council, E/CN.4/Sub.2/2004/30, 8.

8. Daes, Preliminary Report, 17.

9. United Nations, *Draft Declaration on the Rights of Indigenous Peoples* (E/CN.4/Sub.2/1994/2/Add.1 of April 20, 1994, Article 42). At the UN Commission on Human Rights, Indigenous peoples have developed and lobbied for the approval of the Draft Declaration on the Rights of Indigenous Peoples, which has not yet been adopted because of opposition primarily by developed countries such as the United States, Canada, Australia, and New Zealand.

10. International Labour Organization, *Convention No. 169 Concerning Indigenous and Tribal Peoples in Independent Countries* (adopted June 27, 1989, coming into force September 5, 1991).

11. Inter-American Commission on Human Rights, *Proposed American Declaration on the Rights of Indigenous Peoples* (approved on February 26, 1997, at its 1,333rd session, 95th regular session), AG/RES. 1479 (XXVII-O/97).

12. Sharon Helen Venne, *Our Elders Understand Our Rights: Evolving International Law Regarding Indigenous Rights* (Penticton, B.C., Canada: Theytus Books, 1998), 137.

13. United Nations, *Draft Declaration on the Rights of Indigenous Peoples*, Article 26.

14. United Nations, *Draft Declaration on the Rights of Indigenous Peoples*, Article 29.

15. United Nations Environmental Program, *Convention on Biological Diversity* (entered into force on December 29, 1993 and has 187 Parties as of January 31, 2003). The CBD's website (www.biodiv.org) contains links to the provisions of the Convention, subsequent COP (Conference of the Parties) decisions, and other relevant information.

16. The World Conference of Indigenous Peoples on Territory, Environment and Development, *Kari-Oca Declaration and the Indigenous Peoples' Earth Charter* (May 25–30, 1992).

17. Wynberg, Rachel, "Sharing the Crumbs with the San," Biowatch SA, www.biowatch.org.za/csir-san.htm.

18. United Nations, *Beijing Declaration of Indigenous Women* (issued at the UN Fourth World Conference on Women, Huairou, Beijing, People's Republic of China, 1995, 38 and 43).

19. United States Patent No. 5,397,696.

20. Declaration of Indigenous Peoples of the Western Hemisphere Regarding the Human Genome Diversity Project.

21. Palapala Kulike OKa`Aha Pono Paoakalani Declaration, issued in Waikiki, Oahu, Hawaii, October 3–5, 2003. www.ilio.org.

22. Keith Aoki, "Neocolonialism, Anticommons Property, and Biopiracy in the (Not-So-Brave) New World Order of International Intellectual Property Protection" *Indian Journal of Global Legal Studies* 6 (Fall 1999).

23. The Mataatua Declaration on Cultural and Intellectual Property Rights of Indigenous Peoples (June 1993); see at www.ipcb.org/resolutions/htmls/mataatua.html.

24. Mataatua Declaration, Article 15.1.

25. Mataatua Declaration, Article 8 (j).

26. United Nations, *Beijing Declaration of Indigenous Women*, 39.

27. Declaration of Indigenous Peoples of the Western Hemisphere Regarding the Human Genome Diversity Project.

28. WIPO's primary objectives are to administer treaties on intellectual property laws, assist signatory nations in promulgating intellectual property laws, and harmonize and simplify relevant rules and practices. All relevant documents can be seen at www.wipo.int.

29. Victoria Tauli-Corpuz, *Biodiversity, Traditional Knowledge and Rights of Indigenous Peoples* (Third World Network, 2003), 11–12. The workshop summarized in this report was organized by the Tebtebba Foundation in coordination with the Third World Network and GRAIN and was held in Geneva, Switzerland, July 3–5, 2003. Workshop participants included Indigenous representatives from Africa, South America, Asia, the Pacific, the Arctic, and North America and representatives of international and national NGOs. A few representatives of UN agencies were present as observers.

30. International Indigenous Forum on Biodiversity, *Opening Statement Regarding Item 7, Development of Elements of a Sui Generis System for the Protection of Traditional Knowledge, Innovations and Practices of Indigenous and Local Communities*, Sub-Working Group II, Intercessional Ad Hoc Working Group on Article 8 (j) (December 8, 2003). Document in the possession of the author.

31. For example, the *Treaty to Share the Genetic Commons* was signed by over 325 organizations. See www.foet.org/Treaty.htm for more information.

32. Daes, Final Report, 7–8.

33. Daes, Final Report, 7.

34. For instance, the Cherokee Nation established an Institutional Review Board in 1997, and the Navajo Nation enacted its Health Research Code in 1995. Requests for information can be sent to the Cherokee Nation IRB, PO Box 948, Tahlequah, Oklahoma 74465, and the Navajo HRRB Program, PO Box 1390, Window Rock, AZ 86515.

35. *Statement of Proprietary Rights over All Species on Our Traditional Territory*, St'at'imc Nation, Mount Currie, British Columbia, February 22, 2000. The statement can be seen at www.ubcic.bc.ca/papers.htm.

10

Global Trade and Intellectual Property: Threats to Indigenous Resources

Vandana Shiva

$\mathcal{B}$iodiversity is the very fabric of life—it provides the conditions for life's emergence and maintenance, and the many different ways in which that life is expressed. Biological diversity and cultural diversity are intimately related and interdependent. In fact, biodiversity is the embodiment of centuries of cultural evolution and has shaped the world's cultures, because humans have coevolved with other species in the diverse ecosystems of the world. Both biodiversity and cultural diversity have been threatened by the globalization of an industrial culture based on reductionist knowledge, mechanistic technologies, and the commodification of resources.

Our seeds and plants are sacred to us. In the Andean cultures, which have given the world the gifts of corn, amaranth, potato, and quinoa, seeds are the daughters of *La Pachamama* (Mother Earth) and they are our mothers—*Mamalas*. Potato is *Mamala papa*, corn is *Mamala maize*, amaranth is *Mamala Kiwicha*.

The seed for indigenous farmers is not merely the source of future plants and food; it is the storage place of culture and of history. Seed is the first link in the food chain and the ultimate symbol of food security.

Free exchange of seed among farmers has been the basis of maintaining biodiversity as well as food security. This exchange is based on cooperation and reciprocity. A farmer who wants to exchange seed with another individual generally gives a quantity of seed equal to that received.

Free exchange among farmers goes beyond mere exchange of seeds; it involves exchange of ideas and knowledge, as well as exchange of culture and heritage. It is an accumulation of tradition, including the knowledge of how to work the seed. Farmers gather knowledge about the seeds they want to plant in the future by watching them grow in other farmers' fields. This knowledge is based on cultural, religious, and gastronomic oral traditions, as well as shared information about drought, disease resistance, and pest tolerance. Indigenous communities maintain these and other values accorded to the seed and the plant it produces.

Paddy, for example, has religious significance in most parts of India and is an essential component of most religious festivals. The *Akti* festival in Chattisgarh, a center of diversity of the *Indica* variety of rice, reinforces the many principles of biodiversity conservation. In the south, rice grain is considered auspicious or *Akshata*. It is mixed with *kumkum* and turmeric, and given as a blessing. Other agricultural varieties whose seeds, leaves, or flower form an essential component of religious ceremonies are coconut, betel leaves, arecanut, wheat, finger and little millets, horsegram, blackgram, chickpea, pigeon pea, sesame, sugarcane, jackfruit seed, cardamom, ginger, bananas, and gooseberry.

New seeds are worshipped before they are planted, and new crops are worshipped before being consumed. Festivals before the sowing of seeds, as well as the harvest festival, which are celebrated in the fields, symbolize people's intimacy with biodiversity. For the farmer, the field is the mother, and worshipping the field is a sign of gratitude towards the earth, who, as mother, feeds the millions of life forms who are her children.

Patents on life are the main mechanism for expropriating the biological resources and traditional knowledge of indigenous cultures. These expropriations lead to cultural erosion. They also create poverty and loss of livelihood.

Biodiversity is not just a conservation issue; it is an issue affecting economic survival because it is the means of livelihood and the "means of production" of the poor who have no access to other assets or sources of production. For food and medicine, for energy and fiber, for ceremony and crafts, the poor depend on the wealth of biological resources and on their knowledge and skills related to biodiversity. As biodiversity disappears, the poor are further impoverished and deprived of the health care and nutrition that biodiversity provides.

There exists a very intricate relationship between local communities and biological diversity. Hunting-and-gathering communities use thousands of plants and animals for food, medicine, and shelter. Pastoral, peasant, and fishing communities have also developed the knowledge and skills to obtain a sustainable livelihood from plant and animal diversity, in both wild and domesticated forms, situated on the land, in the rivers, and in the lakes and seas. The life of communities has been enhanced spiritually, culturally, and economically as the communities in turn have enriched the earth's biodiversity.

THE CLASH OF TWO TREATIES—WTO VS. CBD

Intellectual property rights (IPRs) are supposed to be property rights to the products of the mind. If IPR regimes reflected the diversity of traditional knowledge that accounts for creativity and innovation in different societies, they would necessarily have to reflect a triple plurality: of intellectual modes, of property systems, and of systems of combinations.

But IPRs, as implemented by nation-states in response to the Uruguay Round of the General Agreement on Tariffs and Trade (GATT), as framed by the World Trade Organization (WTO) rules, or as imposed unilaterally through Special Clause 301 of the U.S. Trade Act, are a prescription for a monocultural knowledge. They are being used to universalize the U.S. patent regime worldwide. This would inevitably lead to an intellectual and cultural impoverishment, since it would diminish and displace other ways of knowing, other objectives for knowledge creation, and other modes of knowledge sharing.

Indigenous cultures have freely shared their seeds and their medicinal plants with the world. Today, patents and biopiracy threaten this culture of sharing and gift giving. Under the new free trade arrangements of the WTO, the privatization of life and the reduction of living diversity and its parts and processes to tradable commodities have been made legal obligations. The WTO has become a global standard based on the logic and primacy of trade and commerce. Trade without limits and without barriers has been elevated to a supreme right, while protection of living resources, livelihoods, and lifestyles has been reduced to a "barrier to free trade."

Nonetheless, the right and obligation to protect life's diversity and diverse lifestyles are also part of an international, legally binding agreement that was signed at the Earth Summit in Rio in 1992. The treaty for the conservation of biodiversity, the Convention on Biological Diversity (CBD), makes the conservation and sustainable use of living resources an international obligation.

The WTO is in direct conflict with the 1992 Rio Convention on Biological Diversity. The former requires the privatization of life through enforcing patents on life, while the latter requires the protection of biological and cultural diversity. The WTO undermines national sovereignty, while the CBD upholds the principle of national sovereignty. Under the CBD, each state regulates access to its genetic resources and can deny access if it appears harmful to its national interests. Under Article 3, the CBD recognizes the sovereign rights that states have in accordance with the Charter of the United Nations "to exploit their own resources pursuant to their own environmental policies, and the responsibility to ensure that activities within their jurisdiction or control do not cause damage to the environment of the States or of areas beyond the limits of national jurisdiction."[1]

Article 8 (j) recognizes that each state should:

> subject to its national legislation, respect, preserve and maintain the knowledge, innovations and practices of indigenous and local communities embodying traditional lifestyles relevant for the conservation and sustainable use of biological diversity and promote their wider application with the approval and involvement of the holders of such knowledge, innovations and practices and encourage the equitable sharing of the benefits arising from the utilization of such knowledge, innovation and practices.

The CBD acknowledges the role of local farmers and tribes in bioconservation, and obliges states to provide avenues for the protection of farmers' and national rights to biodiversity and indigenous knowledge.[2] Furthermore, it exhorts states to protect and encourage customary use of biological resources in accordance with traditional land practices that are compatible with conservation or sustainable use requirements. Articles 10 (a) and 10 (c) direct the contracting Parties to "Integrate consideration of the conservation and sustainable use of biological resources into national decision making and protect and encourage customary use of biological resources in accordance with traditional cultural practices that are compatible with conservation or sustainable use requirements." In accordance with Article 10 (c): "Contracting parties are obliged to protect and encourage customary use of biological resources in accordance with traditional cultural practices to conserve and sustainably use these resources."

In effect, the WTO protects the commercial rights of global corporations that profit from trade in genes, cells, plants, seeds, animals, and their manipulation through patents and other intellectual property regimes. Since these commercial rights are in conflict with the right of all species and all peoples to survival, the WTO calls into question the sanctity of life and undermines the CBD. Global corporations that today have reconstituted themselves as life-sciences corporations have admitted that they drafted the TRIPS Agreement of the WTO, which gives them the right to own and control living resources. As a representative of Monsanto has stated: "We were the physicians, the diagnostician, the patient—all in one."[3]

BIOPIRACY

The distorted IPR laws globalized by the WTO have led to an epidemic of biopiracy—the patenting of biological resources and traditional knowledge of indigenous peoples. *Biopiracy* refers to the use of intellectual property systems to legitimize the exclusive ownership and control over biological resources and biological products and processes that have been used over centuries in non-industrialized cultures. Patent claims over biodiversity and indigenous knowledge that are based on the innovation, creativity, and genius of the peoples of the Third World are acts of biopiracy. Since a patent is given for invention or discovery, a biopiracy patent denies the innovation embodied in indigenous knowledge. The rush to grant patents and reward invention and discovery has led corporations and governments in the industrialized world to ignore the centuries of cumulative, collective innovation of generations of rural communities.

Biopiracy occurs because of the inadequacy of Western patent systems and the inherent Western bias against other cultures. Western patent systems were designed for monopoly control over exports, not for screening all knowledge systems to exclude existing innovations and establish *prior art* that is manifest in other cultures. Western culture has also suffered from the "Columbian blunder" of the right to plunder by treating other people, their rights, and their knowledge as non-existent. *Terra nullius* has its contemporary equivalent in *bio nullius*—treating biodiversity knowledge as empty of prior creativity and prior rights, and hence available for "ownership" through the claim to "invention."

Among the biopiracy battles we have fought, and won, are those related to neem and basmati neem, the free tree of India.[4] The Persian name for this neem is *Azad Darakth*, and the scientific name *Azadirachta indica* is derived from it. In Indian texts written over 2,000 years ago, neem is mentioned as an air purifier and as a cure for almost all types of human and animal diseases, because of its insect- and pest-repellant and anti-feedant properties. It is used on every farm, in every house, almost every day. The combined cultural, medicinal, and agricultural values of neem have contributed to its widespread distribution and propagation across continents.

The Indians have provided knowledge about neem to the entire world. More than 50,000 neem trees shelter pilgrims on the way to Mecca. India has shared the tree and knowledge of its utilization freely with the world community. The freedom of the diverse species to exist and the freedom of people to exchange knowledge about them is best symbolized in neem. But the free tree of India is no more free; more than ninety patents have been granted on it, including claims by American, Japanese, and German companies.[5] In 1995, the Research Foundation for Science, Technology and Ecology in conjunction with the Green Party in European Parliament, the International Federation of Organic Agriculture Movements (IFOAM), and 200 other associates filed claims challenging neem concerned patents granted to W. R. Grace in the European Patent Office. On May 10, 2000, in a historic victory, the neem patent held by W. R. Grace and USDA was revoked by the European Patent Office.[6]

The Indian subcontinent is the largest producer and exporter of superfine aromatic basmati rice. Basmati evolved over centuries of observation, experimentation, and selection by farmers who developed numerous varieties of the rice to meet various ecological conditions, cooking needs, and tastes. On September 2, 1997, the U.S. Patent and Trademark Office (PTO) granted a patent (No. 5,663,484) on basmati rice lines and grains to a U.S. agribusiness corporation, RiceTec, Inc. Basmati, highly esteemed for its unique aroma and flavor, is one of the most superior varieties of rice grown in India. The twenty patent claims on this invention are quite broad in nature. Patent No. 5,663,484 on basmati appears to be exceptionally broad. It not only defines the scope of the patent to cover "novel Basmati rice lines and grains" but also includes nineteen distinct and separate claims within the one patent. The patent covers genetic lines of the basmati developed by farmers. If the patent protection is enforced, farmers will not be able to grow these varieties developed by them and their forebears without getting permission from and paying royalty to RiceTec. The RiceTec strain possesses the same qualities as our Indian traditional varieties, including long grain, distinct aroma, high yield, and semi-dwarf stature. As the RiceTec line is essentially derived from traditional basmati and cannot be claimed as "novel," it should not be patentable.

RiceTec's basmati also cuts into the Indian and Pakistani export market. The United States is one of the largest importers of Indian basmati. If RiceTec is allowed to market its rice in South Asia under the name *basmati*, and at a price cheaper than the Indian varieties, Indian rice exports will be drastically affected. The potential loss in basmati exports due to the RiceTec patent on basmati will severely impact upon the Indian economy. Our basmati rice exports, valued at US$242 million, which constitute almost three-fourths of total rice exports, represent a lucrative niche market for this country. India exports nearly half a million tons of basmati rice annually, mainly to the Middle East, Europe, and the United States.[7] Our movement against basmati biopiracy was successful in striking down most of the false claims to invention in RiceTec's patent.

There is an epidemic of biopiracy—the patenting of indigenous biodiversity and traditional knowledge by global corporations. First it was neem, then basmati. Now our wheat, our "atta," our "chapatis" have been patented. Conagra, a U.S. agribusiness company, was granted Patent No. 6,098,905 for "atta."[8] In 1996, Unilever/Monsanto was granted a patent (EP 518,577) for claims to have "invented" the use of flour to make traditional kinds of Indian bread such as chapatis.[9] On May 21, 2003, the European Patent Office in Munich granted a patent with the number EP 445,929 and the simple title "plants."[10] The patent holder is Monsanto, better known as the world's largest trader in genetically engineered plants. The patent covers wheat exhibiting a special baking quality of low elasticity. Wheat with such characteristics was originally developed in India. Now Monsanto holds a monopoly on the farming, breeding, and processing of this type of wheat.

Since a patent is an exclusive right based on invention, biopiracy patents rob us of our claim to our scientific and intellectual creativity, by allowing indigenous innovations to be treated as "inventions" of the biopirate. For this reason alone, they need to be challenged. But they also have serious economic consequences. In the short run, a biopiracy patent robs us of markets overseas for our unique products. If these trends are not challenged and if IPR systems are not changed to prevent biopiracy, over time we will be paying royalties for what belongs to us and is necessary for everyday survival of our people.

Biopiracy is both legally and morally wrong. Through the patenting of indigenous knowledge, biopiracy is a double theft because first it fosters theft of creativity and innovation, and secondly, it deprives people of everyday survival from the full use of indigenous biological resources. Over time, the patents can be used to create monopolies and increase the price of everyday products. If there were only one or two cases of such false claims to invention on the basis of biopiracy, they could be called an error. However, biopiracy is an epidemic. The problem is not, as it was made out to be in the case of turmeric, an error made by a patent clerk. The problem is deep and systemic. And it calls for a systemic change.

INTERNATIONAL IPR LAWS NEED TO RESPECT INDIGENOUS RIGHTS AND KNOWLEDGE

International law related to biological resources and traditional knowledge needs to be based on the basic rights and values of indigenous cultures. Indigenous societies do not view their resources and knowledge as private property. Therefore their collective, cumulative innovation and their biological and intellectual commons need to be respected and protected. For this, the WTO Agreement on Trade Related Aspects of Intellectual Property Rights (TRIPS) needs to be changed. The mandatory review of TRIPS, which rich countries have been blocking, must be undertaken. The Africa Group had asserted before the Cancun Ministerial:

> The Africa Group is concerned that the review of Article 27.3(b) of the TRIPs Agreement has not been finalised having started way back in 1999. The Group urges all delegations to positively respond to the instructions to the Council for TRIPs from the fourth Session of the Ministerial Conference in paragraphs 12 and 19 of the Declaration. The deadline of December

2002 within which the review was to be finalised and this reported to the Trade Negotiations Committee (TNC) "for appropriate action" has passed. This should be of concern to all delegations especially given that deadlines in other areas of the work programme have similarly passed without concrete results. The review should continue until all the issues are resolved while at the same identifying the issues on which agreement is available.[11]

The protection of genetic resources and traditional knowledge, particularly those originating from developing countries, is an important means of addressing poverty and is rightly a matter of equity and due recognition for the custodians of the genetic resources and traditional knowledge. It is also a matter of law in the context of protecting cultural rights as well as of preserving the invaluable heritage of humankind that biological diversity and traditional knowledge constitute.

Any protection of genetic resources and traditional knowledge will not be effective unless and until international mechanisms are found and established within the framework of the TRIPS Agreement. Other means, such as access contracts and databases for patent examinations, can only be supplementary to such international mechanisms, which must contain an obligation for members collectively and individually to prohibit, and to take measures to prevent, the misappropriation of genetic resources and traditional knowledge.

Patents on life forms are unethical and the TRIPS Agreement should prohibit them, through modifying the requirement to provide for patents on microorganisms and on non-biological and microbiological processes for the production of plants or animals. Such patents are contrary to the moral and cultural norms of many societies that are members of the WTO. Members that view patents on life as contrary to the fabric of their society and culture consider the exception in Article 27.2, for protecting public order and morality, meaningless in this regard.[12]

The Vienna Convention may provide guidance in defining the relationship between the TRIPS Agreement and the Convention on Biological Diversity as well as the International Treaty on Plant Genetic Resources. However, the debate must go beyond this purely legal context and substantively deal with the crux of the issue raised within the framework of the review.

While our governments try to establish our rights in international law, we defend our rights in the footsteps of Gandhi, who said: "As long as the superstition that people should obey unjust laws exists, so long will slavery exist." On March 5, 1999, about 2,000 groups and movements in India joined hands with Navdanya to launch the Bija Satyagraha—the movement to defend people's right to biodiversity, a new freedom movement against the new colonization of life, livelihood, and living resources.[13] *Bija* in Hindi means "seed," and *satyagraha* means "the struggle for truth."

That same day was the anniversary of Gandhi's "Salt Satyagraha." The historic Salt Satyagraha (salt march) was initiated by Gandhi to protect against the colonization of salt as an Indian resource by the Salt Laws imposed by the British Empire. The Bija Satyagraha is a contemporary movement based on non-cooperation with unjust and immoral IPR laws. The Salt Satyagraha was India's refusal to cooperate with the unjust Salt Laws and an expression of India's quest for freedom with equity. The Bija Satyagraha is the refusal to accept the colonization of life through patents. For us, resistance to patents on life and biopiracy is not just a defense of our rights. It is an ethical imperative because other species are our kin. We are all members of the Earth Family—*Vasudhaiva Kutumbkam.*

NOTES

1. United Nations Environmental Program (UNEP), *Convention on Biological Diversity*, 1992.

2. UNEP, *Convention on Biological Diversity*.

3. James Enyart, "A Gatt Intellectual Property Code," *Les Nouvelles*, June 1990, 54–56.

4. Vandana Shiva, *Protect or Plunder* (London: Zed Books, 2002), 57–61.

5. Vandana Shiva, "Enclosure and Recovery of Commons," Research Foundation for Science, Technology and Ecology, Delhi, India (1997), 47–50.

6. Shiva, *Protect or Plunder*, 61.

7. Shiva, *Protect or Plunder*, 56.

8. Navdanya, "Corporate Hijack of Biodiversity," 2003, 24.

9. European Patent Office, Patent No. 518,577.

10. European Patent Office, Patent No. 445,929.

11. Cited in *Bija* 31/32 (Autumn 2003): 40.

12. *Bija*, 31/32.

13. Navdanya, "Campaign Against Biopiracy," 1999.

Indigenous Peoples and Traditional Resource Rights

Graham Dutfield

*I*ndigenous peoples constitute most of the world's cultural diversity and inhabit some of the world's most biologically diverse regions. Many indigenous peoples have accumulated a wealth of knowledge of their surroundings, and this knowledge has been hugely beneficial to the world and to industry. For example, many pharmaceutical products would not have been discovered if it were not for indigenous peoples. Unfortunately, indigenous groups frequently suffer from extreme poverty, ill health, unemployment, lack of access to land and essential resources, and human rights violations. In consequence, cultural diversity is eroding at an accelerating rate.

This chapter presents the case that all indigenous peoples have the right to manage their own biological resources, to preserve their traditional knowledge, and to protect these from expropriation and biopiracy by scientific, corporate, or government interests, and that the international community must protect those rights from transgression by scientific, corporate, or government interests and is obliged to do so on both moral and legal grounds. To present my argument, I divide the rights statement into its constituent clauses.

ALL INDIGENOUS PEOPLES HAVE THE RIGHT . . .

Opponents and skeptics are likely to assume that by starting the statement this way, we are singling out indigenous peoples for special or even privileged treatment as compared with other cultural communities and polities, and even that we are according them a status akin to nation-states. They may even go further by opposing use of the term "indigenous peoples" on at least two grounds.

One of these is that its past use by European colonialists in Africa was derogatory and remains justifiably unpopular in that part of the world. Second, these critics may agree with some Asian and African governments that to differentiate "indigenous" peoples from other citizens is to be unnecessarily divisive and even destabilizing.

My first response is that indigenous peoples are not looking for special treatment. What they want is to be recognized as peoples with a full entitlement to self-determination, just like other ethnic groups and populations with recent experiences as victims of colonialism. International law supports such a claim, which is why some governments are not only unhappy with the word "indigenous" but also dislike the addition of "s" to "people." Despite this, the existence of the International Labour Organization Convention 169 Concerning Indigenous and Tribal Peoples in Independent Countries indicates that the term "indigenous peoples" has standing in international law.[1] As for what rights indigenous peoples are entitled to, Article 1 of the International Covenant on Economic, Social and Cultural Rights makes the situation quite clear: "all peoples have the right of self-determination. By virtue of that right they freely determine their political status and freely pursue their economic, social and cultural development."

As for the applicability of the term, one must acknowledge that there are difficulties when it is applied to peoples outside of countries established by European settlers, such as the United States, Canada, Brazil, Australia, and New Zealand. However, many culturally distinct minority groups in Africa and Asia who live outside the cultural mainstream have suffered similar abuses over the years and may have legitimate rights on the basis of prior occupation of certain areas that are not recognized. In response to those concerned that the term is destabilizing, it is important to be aware that few if any self-designated indigenous peoples are interested in transforming themselves into separate countries on the basis that they are more "indigenous" than the rest of the national population.

TO MANAGE THEIR OWN BIOLOGICAL RESOURCES

Assuming our interpretation is correct that indigenous peoples have a right to self-determination under international law, does it follow that they are entitled to claim biological resources on their territories as their own and to manage these as they see fit? To answer this question, it is important to understand how the international community of nation-states has determined how to assign rights and responsibilities over natural resources.

It is a fundamental principle of international law that sovereign countries are "territorial polities . . . with the same standing in international society, which means that the norm of non-intervention is central—no sovereign has the right to intervene in the affairs of another."[2] It follows from this that independent states have a right to permanent sovereignty over their territories, including the natural resources existing within them. The United Nations General Assembly confirmed this in 1962 by adopting Resolution 1803 on Permanent Sovereignty over Natural Resources.

Another principle of international law with implications for access to natural resources and rights to benefit from their use is that of the common heritage of humankind. This is normally applied to extraterritorial (and extraterrestrial) non-living resources. Common heritage resources are unowned, but members of the international community of nations have an equal duty to conserve them and an equal right to benefit from their use. It is commonly but erroneously supposed that until the 1990s, when the Convention on Biological Diversity came into force, biological and genetic resources were subject to the common heritage principle. It is true that the 1983 International Undertaking on Plant Genetic Resources (IUPGR) of the Food and Agriculture Organization of the United Nations treated plant genetic resources as the common heritage of humankind.[3] But this was not a legally binding agreement and did not really constitute an explicit renunciation by signatory countries of their permanent sovereign rights over natural resources. Evidently, the permanent sovereignty principle always had the upper hand over common heritage for biological and genetic resources, and no longer has any serious competitor except perhaps for intellectual property rights.

What does this imply for the rights of indigenous peoples to manage their biological resources? It follows that if indigenous peoples have a right to self-determination, they also have the right to manage their own biological resources, albeit with certain duties. The above-mentioned International Covenant supports this, entitling "all peoples, for their own ends," to "freely dispose of their natural wealth and resources without prejudice to any obligations arising out of international economic co-operation, based upon the principle of mutual benefit, and international law. In no case may a people be deprived of its own means of subsistence."[4] Is there a conflict with permanent sovereignty? It really depends on how countries implement the principle. Sovereignty implies the existence of a right to exclude other nation-states and foreign entities such as corporations. It is not the same as ownership and can certainly accommodate the ownership rights of non-governmental domestic entities such as private landowners and indigenous peoples, for example. Therefore, there is no necessary conflict with permanent sovereignty. Governments have a legal and moral duty to ensure that no such conflicts arise in practice that would deprive indigenous peoples of their own means of subsistence or usurp their property rights over resources.

TO PRESERVE THEIR TRADITIONAL KNOWLEDGE

It is estimated that 90 percent of the world's languages will be lost by the end of the twenty-first century.[5] Most of the losses will be borne by indigenous peoples, who contribute most of the world's linguistic and cultural diversity. Since so much knowledge is encoded in language, a great deal of traditional knowledge will follow these languages into oblivion. Of course all cultures change over time—old knowledge is lost when it is no longer useful, and is replaced by new knowledge, which may be generated internally or acquired or adapted from elsewhere. Nonetheless, the rapid cultural change experienced by many indigenous groups around the world is often the result of coercive or paternalistic policies that leave communities demoralized and culturally impoverished.

It does not necessarily follow that all members of indigenous groups are interested in preserving traditional knowledge. Neither is it realistic to expect indigenous communities to preserve their knowledge if they see no advantage in doing so. Nonetheless, the cultural rights to which indigenous peoples are entitled should certainly include the right to preserve their traditional knowledge according to their own needs and interests. Lyndel Prott, formerly of UNESCO, identified a set of individual and collective rights that could be described as "cultural rights," and which are supported to a greater or lesser extent by international law.[6] Of these, the following stand out in light of the present discussion: (1) the right to protection of artistic, literary, and scientific works; (2) the right to have one's cultural identity respected; (3) the right of minority peoples to be respected for their identity, traditions, language, and cultural heritage; and (4) the right of people to be protected against having an alien culture imposed on them.[7]

Given the rapid erosion of indigenous cultures, measures to protect indigenous knowledge and the rights of the holders, custodians, and communities need to be implemented with some urgency, and not just for indigenous peoples but for all our sakes. As the late Darrell Posey so poignantly expressed it: "with the extinction of each indigenous group, the world loses millennia of accumulated knowledge about life in and adaptation to tropical ecosystems. This priceless information is forfeited with hardly a blink of the eye: the march of development cannot wait long enough to even find out what it is about to destroy."[8]

TO PROTECT THESE FROM EXPROPRIATION AND BIOPIRACY BY SCIENTIFIC, CORPORATE, OR GOVERNMENT INTERESTS

In August 2001, the Sub-Commission on the Promotion and Protection of Human Rights of the United Nations Commission on Human Rights adopted a resolution on "Intellectual Property Rights and Human Rights," which stressed "the need for adequate protection of the traditional knowledge and cultural values of indigenous peoples and, in particular, for adequate protection against 'bio-piracy' and the reduction of indigenous communities' control over their own genetic and natural resources and cultural values."[9]

Over the years, the knowledge of indigenous peoples has provided valuable leads for the development of new medicines and other industrial products. Consequently, a great deal of wealth has been extracted from their knowledge and resources. Most recently, scientists and companies have become interested in genetic information gleaned from the bodies of indigenous people. Rarely have indigenous peoples benefited materially or otherwise from this interest, such as by receiving a reasonable share of the profits arising from commercialization of their knowledge and resources. To make the situation even more unjust, indigenous peoples often find that companies acquire patent monopolies claiming discoveries that seem hardly different from the knowledge and resources provided to them by the indigenous peoples.

Unfortunately, the modern patent system is unbalanced in that it recognizes certain forms of inventive activity, such as the outputs from corporate research laboratories, but not others. For example, in most cases indigenous peoples' knowledge cannot easily be reduced to a patent specification even if it is just as useful or innovative as a corporate invention. Thus, scientists have been able to patent certain compounds found in a plant called hoodia, which has traditionally been used by certain groups of Bushmen people as an appetite suppressant. But the indigenous groups that showed the scientists how to use the plant were in no position to assert property claims to this knowledge through the intellectual property systems or otherwise.[10] This is unfair both for these peoples and for those developing countries where the presence of local populations possessing and producing such knowledge could potentially provide competitive advantages for their economies. As James Boyle so nicely puts it, a consequence of lack of balance (in the IP system generally) is that "curare, batik, myths and the dance 'lambada' flow out of developing countries, unprotected by intellectual property rights, while Prozac, Levis, Grisham and the *movie* 'lambada!' flow in—protected by a suite of intellectual property laws, which in turn are backed by trade sanctions."[11]

It would go too far to say that patents cannot benefit indigenous peoples. For example, where an indigenous group and a drug company have instituted benefit-sharing arrangements, the latter will need to file patents in order to secure returns from the heavy costs of bringing new products to market. Indeed, the groups which told scientists about their use of hoodia will now receive benefits from the patent-owning institution. Nonetheless, few such arrangements exist, and since patents are essentially "winner-take-all" rights, which enable one party to monopolize all the benefits, the indigenous peoples usually get exploited.

To make matters worse, the patent system nowadays seems to require little in the way of improvement to naturally occurring resources for them to become protectable inventions. Apart from being questionable public policy, this situation encourages misappropriation of indigenous peoples' knowledge. This is especially likely given that many businesses can be very aggressive in claiming ownership of resources they claim to have discovered or invented and in defending their rights once these have been granted.[12]

In conclusion, Article 4 of the Genetic Bill of Rights has ample justification on legal and moral grounds, as well, arguably, as on utilitarian grounds. The problem then arises of how these rights should be put into effect. All of them are clearly enunciated in the draft United Nations Declaration on the Rights of Indigenous Peoples, which is now over ten years old. Unfortunately, the failure of the United Nations member states to adopt an agreed-upon text suggests a lack of will on the part of the international community to recognize its moral and legal duties in respect of the rights of indigenous peoples. While a small number of countries, most notably Peru and Panama, have introduced legal regimes to protect the knowledge of indigenous peoples, the overwhelming majority of governments still do not appear to take this issue seriously.

NOTES

1. Article 1.1 (b) refers to "peoples in independent countries who are regarded as indigenous on account of their descent from the populations which inhabited the country, or a geographical region to which the country belongs, at the time of conquest or colonisation or the establishment of present state boundaries and who, irrespective of their legal status, retain some or all of their own social, economic, cultural and political institutions."

2. Chris Brown, *Sovereignty, Rights and Justice: International Political Theory Today* (Cambridge, UK: Polity Press, 2002), 35.

3. According to Article 1, "This Undertaking is based on the universally accepted principle that plant genetic resources are a heritage of mankind and consequently should be available without restriction."

4. Article 1.2.

5. Michael Krauss, "The World's Languages in Crisis," *Language* 68, no. 1 (1992): 4–10.

6. Lyndel V. Prott, "Cultural Rights as Peoples' Rights in International Law," in J. Crawford, ed., *The Rights of Peoples* (Oxford: Clarendon Press, 1988), 93–106.

7. It is noteworthy that Article 8 (j) of the Convention on Biological Diversity requires state parties, "as far as possible and as appropriate," to "respect, preserve and maintain knowledge, innovations and practices of indigenous and local communities embodying traditional lifestyles relevant for the conservation and sustainable use of biological diversity and promote their wider application with the approval and involvement of the holders of such knowledge, innovations and practices and encourage the equitable sharing of the benefits arising from the utilization of such knowledge, innovations and practices." It must be said that the language is generally vague and it is difficult to ascertain the specific legal requirements—if any—of state parties.

8. Darrel A. Posey, "Indigenous Knowledge and Development: An Ideological Bridge to the Future," in K. Plenderleith, ed., *Kayapó Ethnoecology and Culture* (New York: Routledge, 2002), 59.

9. United Nations Commission on Human Rights, Sub-Commission on the Promotion and Protection of Human Rights, "Intellectual Property and Human Rights," Resolution 2001/21 (E/CN.4/SUB.2/RES/2001/21), 2001.

10. Graham Dutfield, "Sharing the Benefits of Biodiversity: Is There a Role for the Patent System?" *Journal of World Intellectual Property* 5, no. 6 (2002): 906–7.

11. James Boyle, *Shamans, Software and Spleens: Law and the Construction of the Information Society* (Cambridge, MA: Harvard University Press, 1996), 125. For a critical review of commonly articulated concerns about biopiracy, see Graham Dutfield, *Intellectual Property, Biogenetic Resources and Traditional Knowledge* (London: Earthscan, 2004).

12. For example, in 1999, a U.S. patent was granted for a field bean cultivar dubbed "Enola" by its "inventor" Larry Proctor. Proctor's company, Pod-Ners, has been using the patent to block the sale of imported beans with the same color as the one described in the patent, whose description would cover several traditional bean varieties. The patent claims not only a certain yellow-colored *Phaseolus vulgaris* bean seed, plants produced by growing the seed, as well as all other plants with the same physiological and morphological characteristics, but also the breeding methods employed. Soon after receiving the patent, Proctor sued a company called Tutuli that had been importing Mexican yellow bean cultivars called mayocoba and peruano from that country since 1994. With customs inspectors disrupting supplies, Tutuli began to suffer financially, as did Mexican farmers who had been selling their beans to this firm. Proctor's company has filed lawsuits against various other small bean companies and farmers. Dutfield, *Intellectual Property, Biogenetic Resources and Traditional Knowledge*, 54–55.

Part V

ENVIRONMENTAL GENOTOXINS

"All people have the right to protection from toxins, other contaminants, or actions that can harm their genetic makeup and that of their offspring."

Article 5 of the Genetic Bill of Rights

· *12* ·

Arguing for a Right to Genetic Integrity

Marc Lappé

The greatest threat to the integrity of our genetic legacy remains the continuous proliferation of gene-damaging chemicals and their capacity to impede our ability to repair the resulting damage. DNA has been under a barrage of gene-damaging events for eons, notably from background radiation and "natural" genetic damage through error and faulty repair. Against this pattern of entropy, natural selection has ensured that a constellation of repair systems has evolved to redress the bulk of natural mutation and damage. But the addition of a plethora of some 85,000 human-made agents now threatens to tip the scales in favor of damage and loss by increasing mutation rates and interfering with these natural repair systems.

New chemicals enter commerce with little or no evaluation for their genetic toxicity. As yet unevaluated additions to this roster threaten to amplify the pace of genetic damage. At-risk persons are rarely warned about DNA-damaging toxicants until well after exposure has occurred. This pattern has proven true for such widely used chemicals as ethylene oxide, benzidine dyes, and phthalates, which are likely to have caused genetic damage and birth defects in offspring of hospital workers, hair dye users, and plastic formulators.

This advent of more and more highly reactive, DNA-damaging agents into commerce poses a series of underappreciated moral dilemmas. Do persons and their offspring, who do not necessarily benefit materially (if at all) from the proliferation of chemicals and environmental toxicants, have a special claim to protection? Do manufacturers have a special obligation to ensure that gene-damaging chemicals are kept out of general commerce or are at least labeled for their adverse reproductive effects? Do individuals have legitimate claims on producers of gene-damaging chemicals to compensation for any mutational effects?

ASSERTING A RIGHT TO GENETIC INTEGRITY

Persons who assert a right to protection from toxicants that can jeopardize their genetic integrity or that of their offspring must prove at least two things: that the threat is real, and that any potentially exposed person has a valid claim to be protected. The realization that environmental agents could damage genetic material can be traced to the 1920s. During this period, American biologist Hermann Muller did a series of experiments that showed that radiation can produce lethal genetic mutations in fruit flies. At a time when mutations were seen as "natural" genetic variation and the stuff of evolutionary change, Muller championed the idea that mutations create a harmful "genetic load" and that the altered genes could be passed intact from generation to generation. Over time, Muller predicted that natural selection would tend to winnow out the most harmful genes because affected individuals would leave fewer and fewer viable descendants. Unfortunately, he could not have known that subtle genetic damage could in fact build up over time, and that a genetic legacy of damage beyond overt lethal mutations could occur.

Muller was awarded the Nobel Prize for his work on radiation-induced mutations in 1946, one year after the apocalyptic end of World War II. At the time, some in the scientific community voiced a belated concern for the residual damage that might be left by radiation-induced injury. To assess any such damage, the United Nations created an Atomic Bomb Casualty Commission in the 1950s. Headed by geneticist Dr. James Neel, the commission was charged with discerning how much damage was produced to individuals who were born after the atomic and neutron bombs exploded over Hiroshima and Nagasaki. Neel's initial results, which were based on examining data from over 76,000 children, were initially reassuring to a worried world.

Neel and his colleagues reported no overt evidence of birth defects or other damage from nuclear radiation after the birth of children exposed to radiation while in the womb. Women who were exposed while pregnant did have higher than expected rates of miscarriages and children with microcephaly (small brain size). But Neel downplayed these results since widespread hunger and malnutrition followed the A-bomb attacks. In 1985, Johns Hopkins human geneticist James Crow reevaluated the Hiroshima and Nagasaki data, and could still find few links between a person's exposure to a mutagen and the likelihood of overt birth defects in her offspring.[1]

Population geneticists now suspect that Neel and Crow used too broad a brush in evaluating mutations, and that further genetic damage was hidden from their view. As a follow-up to their pioneering work, Neel and his colleague William Schull examined the possibility that damage might have been expressed only in future generations. Data from 9,122 pregnancies that were conceived *after* August 1945 showed that mortality in childhood rose to 64 and 77 per 1,000 births in Hiroshima and Nagasaki respectively, over 10 times higher than the expected mortality rates. In 1965, Neel and Schull reexamined the atomic survivor data, reasoning that any recessive lethal combinations would be more likely to show up in first-cousin marriages. They found that mortality in offspring of first-cousin marriages among A-bomb survivors was even higher, confirming the presence of hidden genetic damage.[2]

HOW SERIOUS IS THE RISK?

From the 1960s to 1980s, the ambiguous data from these early studies perpetuated the notion that "a few mutations can't hurt you." No full-scale program directed at uncovering the extent of the risk posed to genetic integrity from radiation or gene-damaging chemicals was evident until the 1990s. By then, a growing body of evidence implicated genetic instability and mutation as playing a pivotal role in producing cancer and birth defects. About 20 percent of all birth defects are due to gene mutations, another 5–10 percent to chromosomal abnormalities, while perhaps another 10 percent are due to direct physical or chemical damage from pollutants outside the genetic makeup of the cell.[3]

The corollary to this observation is that if we allow the number of chemical mutagens to increase in the environment, we will increase the risk that they will disrupt sensitive genetic machinery. Researchers now recognize that the concerns voiced by Muller for radiation-induced mutational damage were prescient. In 2002, a group of British and Russian researchers found that genetic damage produced by irradiating animals was passed to their offspring in the form of a generalized genetic instability, not simply one damaged gene at a time. After a single radiation exposure, the overall rate of mutation in two subsequent generations of mice was increased significantly. These findings suggest that gene-damaging events can destabilize the genome, producing reverberating damage over multiple generations. The authors flagged the significance of their study with a single sentence: "The remarkable finding that radiation-induced germline instability persists for at least two generations raises important issues of risk evaluation in humans."[4]

If true, this finding doubles the level of concern about the genetic consequences from exposure to radiation and radiation-mimicking chemicals. Many of the most gene-damaging chemicals, such as ethylene oxide, are known as "radiomimetic" (literally mimicking the effects of radiation, specifically producing gene deletions, base-pair substitutions, and chromosome breaks) in their adverse effects. As a leader of a team of scientists from the Swiss Federal

Institute of Technology in Zurich observed, such "environmental genotoxins [DNA-damaging chemicals] can directly alter gene pools."[5]

Genetic changes also lead to cellular disruption and eventually to cancer. As tumors "progress" to freedom from the normal constraints on cell division, they accrue more and more mutational events and develop chromosomal abnormalities. Eventually entire chromosomes break and rearrange, giving rise to "aneuploidy," in which the chromosome count itself is deranged. Any further exposure to a chemical after an initial cancer-producing event will thus exacerbate the damage.

The message from the cancer cell is that protecting the integrity of the cell requires stable and intact genetic material. Control of genetic damage is so critical to natural cell function and longevity that almost all organisms have devised a wide range of quality control, including excision and repair enzymes to ensure the integrity of their DNA. When those repair systems themselves are damaged, as can occur from exposure to chemical mutagens at the level of the germline, cancer is even more likely. Mutations in the BRCA1 and BRCA2 genes or in the TP53 genes involved in DNA repair create high risks of cancer in tissue such as that in the ovaries or breast, or solid organs in the body more generally.

CLAIMING A RIGHT FOR PROTECTION

People can reasonably assert the right to protection from gene-damaging chemicals on at least three grounds: (1) they have a right to be free of non-consenting exposure to agents that create undue risks to their genetic integrity; (2) they have a fundamental right to procreate without fear of passing genetic damage to their offspring; and (3) it is in society's general interest that persons pass on an intact genetic legacy to future generations.

The first claim is predicated on the general privacy rights enjoyed in the United States and elsewhere. By analogy to certain provisions of property law, persons can reasonably assert their right against chemical trespass. The second claim is validated by analogy to the general success of medical malpractice and toxic torts for wrongful life in the United States. Where a company has been found guilty of negligence in allowing a harmful chemical into commerce that produces birth defects, it is required to compensate the child or its heirs. For instance, the classic case of *Sindell v. Abbott Laboratories* was considered by the appellate court to be so critical, it decided in favor of the plaintiff (who suffered from diethylstilbestrol-induced cancer while in her mother's womb) even though the mother could not ascertain which company sold her the drug at issue.

The third claim of a general obligation not to imperil future generations is embedded in common law and governmental policies to protect the environment, such as the National Environmental Protection Act. The lesson of six decades of pollution is that, no less than intact forests and oceans, genomes are a legacy that will deteriorate if not protected. Contemporary studies, such as those cited above, indicate that a significant degree of risk from genetic damage is likely to continue in future generations. Over time, human beings remain at some as yet indeterminate risk for serious health consequences from any past increase in their mutational load. The claim for protection from chemically induced genetic harms is thus analogous to both the right of offspring to seek redress after toxic injury and the privacy right to be free from ongoing chemical trespass embodied in property law.

DUTIES CONFERRED BY RIGHTS

These assertions of rights create a complementary set of duties for others to vouchsafe genetic privacy and unwarranted and non-consensual damage to personal genetic material. A case in point would be ozone-depleting chemicals. Ozone depletion, which permits a tremendous flux of UV light to enter the earth's biosphere—and with it, the risk of millions of new mutations per cell per day—is largely caused by specific chlorine-containing molecules. The right of people to be free from toxicants that can jeopardize their genetic makeup was part of the argument that led to the 1976 decision in California to ban the worst ozone-depleting offenders, the chlorinated fluorocarbons (CFCs) used as propellants in aerosol cans.[6] The first duty to rein in both direct and indirect gene-damaging chemicals such as the CFCs has been recognized in international treaties such as the Kyoto Protocol.

A second duty is to provide the monitoring or surveillance needed to minimize the impact of any mutagenic chemicals in commerce. This duty was advocated in 1993 by Professor F. H. Sobels of the Department of Radiation Genetics and Chemical Mutagenesis of Leiden University. Sobels told an international forum of scientists that an effort to assess and quantify genetic risks from human exposure to mutagenic chemicals was "urgently needed."[7]

To this end, the U.S. Environmental Protection Agency created the so-called Gene-Tox program. Its mission to identify environmental mutagens initially focused on chemicals that had proven to be carcinogens and on assessing what portion carried genetic toxicity. Their results generally confirmed the high correspondence (approximately 85 percent) between carcinogens and genetic toxins, at least in rodents.[8] Gene-Tox scientists also studied non-carcinogens in animals, but little or no data on genetic toxicity were available. Through their efforts, a cadre of several chemicals (out of hundreds of candidates) that were retested were found to be gene-damaging.[9]

A third duty is a corollary of the second: to chart the extent and nature of potential human exposure to mutagenic chemicals and any special toxicological sensitivity that exists in the at-risk population. This work will be invigorated by an infusion of research support and funds into a newly created "Comparative Toxicogenomics Database" being developed at the Mount Desert Island Biological Laboratory in Salisbury Cove, Maine. This new program will concentrate on acquiring information about gene sequences changed or altered by toxic agents, as well as assembling a database of information about the genes and proteins associated with toxicological phenomena.[10]

A fourth duty is to warn about genotoxic risks. This goal is partially achieved by California's Proposition 65 rules requiring that notification and warning signs accompany any consumer product to which a daily exposure produces a significant risk of reproductive or developmental harm. Under the Safe Drinking Water and Toxic Enforcement Act of 1986, companies are responsible for labeling any risk-generating product under a set of rules of exposure.

The fifth duty is to compensate those who have already been injured by gene-damaging events. Identified persons would not only be those injured in wartime, but claimants such as the DES daughters who assert harm through collective exposure to untested or uncontrolled use of gene-damaging chemicals.

A CASE STUDY

A widely used chemical group provides a model for justifying limits on gene-damaging chemicals. Phthalates are used to soften plastics, to improve personal care products, and to increase the solubility of other chemicals in paints, glues, and insect repellents. In 1987, researchers found that the phthalate known as DEHP (for diethylhexylphthalate) and its metabolites were discovered in patients on dialysis or receiving multiple blood transfusions. Since then, four different phthalates have been consistently found in the urine samples collected by the Third National Health and Nutrition Examination Survey designed to gather information about the U.S. population.

One of these chemicals, known as monoethyl phthalate (MEP), is associated with increased DNA damage in sperm. In 2003, researchers reported that men with increasing environmental exposure to MEP had proportional increments in the number of breaks in their sperm DNA.[11] Given the prevalence of MEP in some percent of the population, this result suggests that the integrity of DNA may be compromised across a broad swath of American men. As yet, there is no specific legal recourse for genetic damage per se, without evidence of medical harm. In the interim, taking the precautions of requiring labeling, alerting producers to the risks posed by their reliance on phthalates, and finding alternatives appear highly desirable. In this vein, Baxter Healthcare Products has recently found substitutes for the phthalates in their blood and dialysis product lines.

FUTURE PERSPECTIVE

While clearly concerned about mutagenic damage, the scientific establishment in the United States has not yet embraced a sense of urgency for protecting the genetic integrity of either individuals or the population as a whole. The assertion that all persons have an equal claim on the integrity of the genome because it is part of a "common heritage" was rejected by the team assembled by the American Association for the Advancement of Science (AAAS) to examine the moral and ethical implications of the Human Genome Project.[12] A similar position that persons have a right to an unaltered genetic patrimony, adopted by the Parliamentary Assembly of the Council of Europe,[13] was also rejected.

The AAAS group denied that the human species has a common germ-line. In their words, "The human gene pool is a heuristic abstract, not a natural object, and lacks a material referent in nature. Individuals inherit a specific set of genes derived from their parents. Thus from a biomedical perspective, there is no intergenerational 'human germ line' that could serve as an asset to the future."[14] In this way, the AAAS research group rejected the idea that we have any common responsibility for the genes carried by our species.

The facts of familial inheritance aside, this view is ideologically dangerous in the extreme. If we have no collective resource deserving of protection, we cannot be held responsible for the collective genetic damage now occurring silently from environmental contamination. The Human Genome Project specifically, and the EPA in particular, should be held accountable for their neglect of the tremendous potential for intergenerational harm that can result from ignoring genetic warning signs. Presently, in spite of the small Gene-Tox program in the EPA, we are allowing thousands of new gene-damaging chemicals to enter the biosphere each year.

The resulting proliferation of genotoxic chemicals is largely invisible and unrecognized. Even a list of "reproductive toxicants" falls short of identifying the worst offenders because targeted chemicals are merely "toxic" to the endocrine or germ-cell producing system. The phthalate chemical cited above, which was found to directly damage sperm DNA, is an exception to the rule.

When chemicals or elements like lead, ethylene oxide, MTBE, or chlorinated solvents are found to produce "testicular damage," that effect is simply added to organ-specific toxicity generally. Any secondary effect on the genetic quality of sperm has been largely ignored. How many chemically exposed, subfertile men have passed on genes that have been damaged or mutated? The research literature includes at least one example of a man with a small percentage of a specific form of damaged sperm who passed his abnormality on to his two male children, both of whom were infertile.[15]

While a few investigations concerning genetic damage and human fertility are taking place on the level of the individual,[16] few if any studies are effectively following the leads of environmental scientists who point to broader concerns, such as the decline in human fertility in the general population over the past few decades. It is precisely this disconnect between personalized medicine, individual gene therapeutics, and the global oversight about what is happening to the genetic quality of the human species or the biosphere more generally that is most disturbing.

CONCLUSION

Future generations have a right to "an unmodified" human genetic inheritance, not simply one that protects the gene pool from inadvertent or intentional germline genetic engineering. More particularly, while the public cannot claim a right of protection from the inevitable "natural" load of genetic mutations, it can claim a right against genetic damage from human-made activities. Asserting this right recognizes the claim of future generations not to be left worse off than we are by being burdened with irreversible damage to our genetic resources. This right to freedom from toxic harm encompasses our present and future genetic makeup and its corollary, the protection of genetic resources by preventing damage to genes or to their repair mechanisms (e.g., through toxic damage by UV light or heavy metals). This right ultimately acknowledges the obligations that come from the duty to protect our common genetic heritage from damage.

The resulting right to an untrammeled, intact genome is jeopardized by the proliferation of chemicals without regard to their gene-damaging properties. To reduce the likelihood of harm, it is imperative that new rules be put in place that screen chemicals and disqualify those for general use that pose the most severe threat of harm. Finally, justice requires that we compensate those whose genetic future has been compromised by non-consenting exposure to gene-damaging chemicals.

NOTES

1. See J. F. Crow and C. Denniston, "Mutation in Human Populations," *Advances in Human Genetics* 14 (1985): 59–121.

2. W. J. Schull and J. V. Neel, *The Effects of Inbreeding on Japanese Children* (New York: Harper and Row, 1965).

3. F. E. Wurgler and P. G. Kramers, "Environmental Effects of Genotoxins," *Mutagenesis* 7 (1992): 321–7. See D. A. Beckman and R. L. Brent, "Mechanisms of Teratogenesis," *Annual Review of Pharmacology and Toxicology* 24 (1984): 483–500.

4. R. Barber, M. A. Lumb, E. Bouton, I. Roux, and Y. E. Dubrova, "Elevated Mutation Rates in the Germ Line of First- and Second-Generation Offspring of Irradiated Male Mice," *Proceedings of the National Academy of Sciences USA* 10.1073/pnas.102015399 (May 7, 2002).

5. Wurgler and Kramers, "Environmental Effects of Genotoxins."

6. M. Lappé, personal observations and discussions with the California State Legislature during the 1976–1977 legislative session.

7. F. H. Sobels, "Approaches to Assessing Genetic Risks from Exposure to Chemicals," *Environmental Health Perspectives* 110, suppl. 3 (2003): 327–32.

8. S. Nesnow and H. Bergman, "An Analysis of the Gene-Tox Carcinogen Database," *Mutation Research* 205 (1988): 237–53.

9. M. D. Waters, H. B. Berman, and S. Nesnow, "The Genetic Toxicology of Gene-Tox Non-Carcinogens," *Mutation Research* 205 (1988): 139–82.

10. See in particular the commentary by C. J. Mattingly, G. T. Colby, J. N. Forrest, and J. L. Boyer, "The Comparative Toxicogenomics Database (CTD)," *Environmental Health Perspectives* 111 (2003): 793–5.

11. S. M. Duty, N. P. Singh, M. J. Silva, et al., "The Relationship Between Environmental Exposures to Phthalates and DNA Damage in Human Sperm Using the Neutral Comet Assay," *Environmental Health Perspectives* 111 (2003): 1164–69.

12. E. Agius, "Germ Line Cells: Our Responsibilities for Future Generations," in S. Busuttil, E. Agius, P. S. Inglott, and T. Macelli, eds., *Our Responsibilities Towards Future Generations* (Valetta, Malta: Foundation for International Studies, 1990), 133–43.

13. Parliamentary Assembly, Council of Europe, Recommendation 934 on Genetic Engineering, adopted January 26, 1982.

14. M. S. Frankel and A. R. Chapman, *Human Inheritable Genetic Modifications* (Washington, DC: American Association for the Advancement of Science, September 2000).

15. S. Florke-Gerloff, E. Topfer-Petersen, W. Muller-Esterl, et al., "Biochemical and Genetic Investigation of Round-Headed Spermatozoa in Infertile Men Including Two Brothers and Their Father," *Andrologica* 16 (1984): 187–202.

16. See D. Escalier, "What Are the Germ Cell Phenotypes from Infertile Men Telling Us About Spermatogenesis?" *Histology and Histopathology* 14 (1999): 959–71.

• *13* •

Refocusing Genomics Toward the Human Health Effects of Chemically Induced Mutations

Sheldon Krimsky

If one were to use the media as a guide, it appears that the vast majority of public and private resources invested in human genetics has been dedicated to studying inherited diseases. We have had articles on longevity genes, cancer susceptibility genes (skin, breast, colon, and prostate), genetic clues to Alzheimer's disease, and genes for obesity, burning up fat, retina degeneration, deafness, food choices and sensitivity, and even some forms of bed-wetting. Many of these findings have come about through studies of families or of disease groups. The implication one gets from reading these reports is that the genetic abnormalities reported are part of life's lottery. Our so-called faulty genes are passed on to us from our parents, their parents, and so forth. The Human Genome Project reports that there are about 4,000 hereditary diseases that are caused by a defect in a single gene.[1]

Two types of genetic abnormalities do not originate from our parents. The first are the non-inherited genetic mutations that occur in utero and that result in some developmental abnormalities. Whether spontaneous or from an environmental factor (transplacental mutagens or developmental genotoxins), the outcome of these mutations may be a genetic abnormality that is not inherited but could be heritable, if they affect the germ cells of the newborn. Since the discovery of transplacental carcinogenesis in the mid-twentieth century, greater attention has been given to the body burden of toxic chemicals in pregnant women. It is believed that some adult and pre-adolescent cancers may have been activated in fetal development by carcinogens in the uterine environment—for example, the synthetic hormone diethylstilbestrol (DES), administered to millions of pregnant women, which caused a rare form of cancer in the adult daughters of these women.

The second type of abnormality occurs post-natally and is the result of somatic or germ-cell DNA alterations caused by environmental factors or that arise from spontaneous mutations of unknown origin. Some causes of somatic mutations are well known. They include the natural radiation of the sun, radioactivity, viruses and bacteria, and mutagenic chemicals. While these mutations are generally not heritable, they can be when the testes or ovaries have been the targets of mutagens.

There are many varieties of mutations, some of which do not result in a disease or phenotypic abnormality. For example, in one biological dictionary, a "silent mutation" is defined as a mutation that causes no noticeable change in the biological activity of the proteins a gene codes for.[2] Also, a spontaneous mutation is defined as a mutation that occurs without being affected by a mutagen.

There are reasons why private and most public sources of funding for genomics seem to play down the non-inherited diseases and the non-inherited genetic mutations. There is little money to be made in such discoveries and lots of money to be lost. The costs on the corporate profits would be substantial if all mutagenic agents, which have become an accepted part of industrial processes, were banned or severely restricted.

Given that somatic and germ-cell mutations can occur spontaneously, result from natural substances such as radiation, food components, and microorganisms, or be caused by synthetic mutagens, the concept of a protected genome seems illusory. Scientists have argued that the spontaneous somatic mutation rate is too low to account for the rate of mutations observed in human tumors. Our bodies may have evolved protective mechanisms to deal with specific types or a certain rate of mutations occurring among selected DNA sequences. What happens when the rate of mutations is enhanced significantly by exposure to human-made substances? What happens when the new synthetic chemicals target a group of cells for which we lack the repair mechanisms?

It is generally recognized that DNA is highly susceptible to environmental insults that cause mutations (alterations in the DNA sequence). The modern concept of the genome speaks of a highly dynamic system undergoing a process of mutation and repair. Biologists study the role of proteins and enzymes in DNA repair mechanisms. These repair mechanisms may function before or after DNA replication.

Some mutations are believed to affect the cell's regulatory capacity, resulting in cells that over-proliferate. Other mutations can block the enzymes that copy DNA (DNA polymerases), resulting in aberrant cells or preventing cells from proliferation (cell death).

The connections between mutations and disease have provided a lively area of research activity and controversy. At first it was thought that there was a one-to-one relationship between mutagenesis and carcinogenesis; that is, mutagens caused cancer. Bruce Ames developed a bacterial assay (Ames Test) in the 1970s to test mutagenesis of chemicals. Then it was thought that such assays could provide a first line of defense against cancer. By screening chemicals for mutagenicity, we could weed out the potential carcinogens. As scientists studied the Ames Test in greater depth, they found the linkage between bacterial mutagenesis and human cancer to be more complex and less deterministic than they had originally believed. Of nearly 500 chemicals studied, it was found that 79 percent of the mutagens were also carcinogenic and 43 percent of the carcinogens were not mutagens. In addition, 55 percent of the noncarcinogens were mutagens.[3] Scientists began looking more closely at DNA repair. The conventionally held view today is that cancer, defined as unregulated cell proliferation, is a multistep process, which is the result of a stepwise accumulation of mutations in certain DNA sequences.[4]

A growing number of scientists also have begun to question the somatic mutation theory of cancer, the dominant theory that underlies cancer research, and believe that cancer is a gross tissue phenomenon as opposed to a mutation effect.[5] According to Sonnenschein and Soto, neoplastic cells are an emergent phenomenon resulting from a flawed interaction between cells and tissues. If mutations were the cause of cancer, they ask, why when cancer cells are taken from one animal and transplanted into another do they revert back to behaving like normal cells?[6] Whichever theory of human carcinogenesis ultimately prevails, the relationship between mutations and disease and the cause of mutations will continue to be an area of significant scientific interest.

For example, the scientific literature asserts that some individuals with alterations in the repair genes may be more prone to mutation-based carcinogenesis. Moreover, some cells may be more prone to malignancies than other cells also because they lack specific repair proteins. This is a widely shared view in what is generally known as the somatic mutation theory of carcinogenesis.

If the conventional view of the relationship between mutation and disease is correct, organisms are in a constant state of DNA alteration and repair. Only when the mutation-repair equilibrium is punctuated by some imbalance will disease result. How then can we view mutations as unnatural or mutagens as xenobiotics? In this analysis, the emphasis is taken away from the mutagens and placed on the repair mechanisms of the organism.

The fact that the relationship between mutagens and carcinogens is more complicated than once believed does not negate the basic scientific finding for many scientists that chemicals that damage DNA are likely to be but are not always carcinogens. The fact that there are components in food that alter human DNA in vitro, and that the food is by all measures safe to humans, should not give us solace that other chemicals that alter DNA in vitro or in vivo are also safe. Our bodies may have evolved to repair DNA alterations caused by certain substances at certain sites in the genome. However, our bodies may not be capable of repairing DNA alterations by xenobiotics that are new to the human species.

Since the development of the Ames Test for mutagenicity, it has been learned that the human diet contains a variety of natural mutagens. These are found in fresh vegetables (plant alkaloids and toxins) and as a result of cooking foods (hetrocyclic amines). For example, charred and fried chicken and beef are considered the most important source of food mutagens. Humans evolved by cooking foods on the open fire. Why should open-fire cooking select for disease? And yet the potency of mutagens from food burnt on the open flame seems to be high.

It would seem that mutagens to which humans are newly exposed are prima facie more suspect than mutagens with which humans have been in contact for centuries. This is hardly a universally held position. At Berkeley, scientists involved with the Human Exposure Rodent Potency Project (HERP) believe that humans are extremely well protected against synthetic chemicals and that most defense enzymes are effective against both natural and synthetic chemicals, including mutagenic agents. They argue that the public has developed irrational fears of synthetic chemicals. Human exposure to natural mutagens is much greater than to synthetic mutagens by orders of magnitude.[7]

If natural chemicals (e.g., those found in the food supply) are the primary human risk from mutagens, then it is difficult to defend a "right" to protection against them. It would be like saying "we have a right to be protected from the sun's UV light." But if among the 86,000 industrial chemicals currently in commerce a non-trivial number are animal mutagens, and human agency, not nature, is the reason we are exposed to them, then the concept of rights can be entertained. Those whose actions with regard to the introduction of toxic chemicals inflict harm on persons bear responsibility to cease such actions to the extent that the afflicted assert their right to be free of such harm. The same right cannot be made of natural substances unless they are transformed and packaged to consumers (e.g., mutagenic herbs, roots, and plant extracts).

Pregnant women exposed to mutagens may pass those through the placenta to the developing fetus. Thus, mutagenic chemicals may affect the adult somatically or may affect the germline of the fetus of a pregnant woman. Mutagens that damage the fetus are generally viewed by society as a more serious assault on human health because the DNA damage may be passed to multiple generations.

When we speak about the right to protections against harm to our genetic makeup from toxins, we are really referring to the chemicals of the industrial age to which human systems may not have adapted. These chemicals are believed to place a burden on the body's DNA repair system. It is a way of stating that we should be testing chemicals for their human mutagenicity before they are released into the environment. The concept of a "right" implies that there is an obligation on some other individual or institution. In this case the obligation must be on the companies that produce new chemicals and the governmental bodies that regulate their use. The highest obligation is to "do no harm." This means "do not introduce harm." The right to the protection against human-made mutagens is consistent with this moral imperative. Unless we assert the right to such protection, human-made mutagens are placed in the same moral category as natural or spontaneous mutagens. It would be like placing automobile accidents in the same moral category as manufacturing flaws, both resulting in human casualties.

Scientists are beginning to develop techniques that will help us distinguish between heritable and non-heritable mutations. For example, scientists at the National Human Genome Research Institute and the National Institutes of Health have developed techniques to distinguish hereditary from non-hereditary tumors. Another group of scientists has been working on mutational spectra—a combination of technologies that will allow researchers to detect and characterize the mutations in populations exposed to mutagenic agents without the need for phenotypic selection or time-consuming DNA sequencing.[8] The vision for research on environmental mutagens has been articulated. It involves a combination of exposure analysis and the use of biomarkers—a method called molecular epidemiology.[9]

The "right" that people have for the protection of their genome (somatic or germline) from environmental mutagens can only be realized when sufficient resources are available to carry through the scientific analysis. It is quite likely that many more diseases are caused by environmental toxicants, including mutagenically induced diseases, than are currently realized from inherited genes.

NOTES

1. See darwin.nmsu.edu/~molbio/diabetes/human.html.

2. Biotech Dictionary, University of Texas, biotech.icmb.utexas.edu/search/dict-search.phtml?title=mutation.

3. Publications from the Carcinogenic Potency Project, potency.berkeley.edu/text/drugmetrev.table3.html.

4. Louise Vander Weyden, Jos Jonkers, and Allan Bradley, "Cancer: Stuck at First Base," *Nature* 419 (September 12, 2002): 127–8.

5. C. Sonnenschein and A. M. Soto, *The Society of Cells: Cancer and the Control of Cell Proliferation* (Oxford, UK: Bios, 1999).

6. Sheldon Krimsky, review of *The Society of Cells: Cancer and the Control of Cell Proliferation*, by C. Sonnenschein and A. M. Soto, *BioScience* 49 (September 1999): 747.

7. Human Exposure Rodent Potency Project, University of California, Berkeley, potency.berkeley.edu/herp.html.

8. William A. Suk, Gwen Collman, and Terri Damstra, "Human Biomonitoring: Research Goals and Needs," *Environmental Health Perspectives* 104 (May 1996): 479–83.

9. Radim J. Srám, "Future Research Directions to Characterize Environmental Mutagens in Highly Polluted Areas," *Environmental Health Perspectives* 104 (May 1996): 603–7.

• *14* •

"Omics," Toxics, and the Public Interest

José F. Morales

Science and technology are key engines in the transformation of societies. One such transformation is the biotechnology revolution. Biotechnology is the diverse set of biology-based techniques whose purpose can vary from generating basic knowledge of living systems to specific commercial applications. The biotechnology revolution is an unprecedented expansion of biological information and techniques affecting many aspects of modern life. In just the last decade, developments in biotechnology have included the sequencing of the human genome, cloning, DNA-based legal decisions, and genetically modified food. These rapidly emerging developments highlight biotechnology's continuing expansion and impact.

However, biotechnology's current social consequences exist in the shadow of its history. In the early twentieth century, premodern genetics not only generated foundational knowledge of heredity but also eugenics and racial hygiene. These disciplines undergirded "hereditarian" movements that led to horrendous crimes in Western societies. In response to these injustices, nations introduced new rights to regulate these social forces.

The United Nation's Universal Declaration of Human Rights is among the world's foremost rights mechanisms. Before the completion of the Human Genome Project, UNESCO extended the scope of human rights to the human genome, announcing the "Universal Declaration on the Human Genome and Human Rights," a seminal human rights document of the fledgling "biotech century." As important as the Declaration is to the world community, the scope of genome-related issues is very broad. Thus, the Genetic Bill of Rights (GBR) and other efforts were developed to expand fully the scope of human rights in response to twentieth-century abuses and in anticipation of the social consequences of the biotechnology revolution.[1]

The "toxics" article (Article 5) of the GBR[2] is a reflection of the modern environmental movement in the United States. This movement arose from earlier public health and conservation movements in response to unbridled industrialization. Industrialization's unprecedented exploitation of natural resources resulted in the exposure of workers to many serious health risks and the disposal of toxic waste products in the land, water, and air. Following nearly a hundred years of accumulating costs of industrialization, the 1970s environmental movement began to redress such hazards by claiming the "right" to clean air, water, and environmental protection. This growing environmental consciousness absorbed earlier developments in radiation and chemical mutagenicity, atomic era radiobiology, and finally the genetic explosion that followed the discovery of DNA structure, which laid the foundation for a new genetic toxicology.

The 1970s saw this accumulated bio-knowledge applied to the risk of disease from toxics. These risk methods drew criticisms from concerned citizens[3] to the scientific community,[4] eventually creating a consensus on methodological inadequacies and a vacuum for science-based risk assessment. In 1990, this vacuum was filled by the Human Genome Project and related DNA research on how sequence variability modulates risk. Approximately five years later, in the dawning age of high-throughput biology, this work converged in the cataloging of variation in environmentally responsive genes[5] and whole genome studies of gene expression.[6] Thus, the "omics" of susceptibilities and genotoxic responses has become the new scientific basis for risk.

THE NEW TECHNOLOGICAL LANDSCAPE

The biotechnology revolution is having a major impact on environmental health.[7] One key impact is an "omics"-based reformulation of risk assessment. "Omics" is shorthand for an array of disparate, yet related, high-throughput techniques deployed in many areas, including environmental health. "Omics" has as its defining aspect high-throughput technology that combines robotics, electronics, miniaturization, software, and hardware to evaluate large numbers of biomolecules in small periods of time. Particular biomolecules have their respective "omics": *genomics* corresponds to DNA sequence; *toxicogenomics* examines RNA responses; *proteomics* studies proteins; *metabonomics* corresponds to small molecules. Other "omics," such as *glycomics*, which studies carbohydrates, exist as well. These "omics" generate great volumes of data that are stored in databases and handled by advanced computational methods also known as *bioinformatics*.

In environmental health, these "omics" are used to study the effects of toxic exposures. These exposures result from human contact with a chemical or physical agent through the skin, lungs, or digestive track. Once in bodily contact, the agent penetrates various cells, is sometimes metabolized, and interacts with DNA. This interaction can produce various effects, including DNA damage. While DNA repair machinery frequently corrects the damage, sometimes DNA is copied incorrectly, resulting in a change in DNA sequence (mutation) or other effects on DNA function. These changes in DNA sequence and function can result in altered cell function and disease. Since "omics" will provide rigorous, human-based information on these toxic effects, it is expected to decrease environmental health uncertainties. Increases in "omics"-based throughput (chemicals tested/unit time), with equally large decreases in cost, may eventually fill in the toxicological data gaps for many environmental chemicals. Several "omic" technologies relevant to studying effects on DNA function are genomics, toxicogenomics, metabonomics, proteomics, and bioinformatics.

Genomics studies the complete DNA sequence of a living organism. DNA sequence is the order of the chemical "letters" making up the DNA molecule. The Human Genome Project (HGP) completed sequencing the full complement of human DNA of a small group of individuals because, early on, it was decided that the HGP wouldn't tackle human genetic variation. Today, genomics studies how variation in the genome underlies differences in traits among individuals. While all humans have the same approximately 30,000 genes, different individuals have variations of these same genes, with unique "spelling" differences in their DNA sequence called SNPs (single nucleotide polymorphisms). When SNPs occur within a gene, the resulting protein's function can vary widely, impacting a host of individual characteristics such as disease susceptibility and the response to medicines, foods, and environmental toxics.

Toxicogenomics utilizes a new high-throughput technology called microarrays or "gene chips" to study the expression of many thousands of a cell's genes at one time. Microarrays are tiny grids of ultra-small dots of fluorescently labeled DNA placed on a small glass or plastic chip by a robot. Each dot corresponds to a gene in the genome. When cells are exposed to an environmental stressor, the "volume" of many genes is turned up or down, thereby producing more or less RNA versions of a gene. These cells are broken open, the RNA is processed, and fluorescent tags are attached and then placed on the chip. Thus, each dot has a certain fluorescence, and the whole grid has a fluorescence pattern depending upon the "volume" or expression of all the genes corresponding to cellular responses to toxic substances. Thus, toxicogenomics uses gene expression to study and understand toxicity in human beings.[8]

Metabonomics examines small biological molecules or "metabolites," the by-products of an organism's metabolism. Many of a cell's proteins convert food into small biological molecules (disassembly lines). Still other proteins are committed to complicated assembly lines that use these metabolites to build complex biological molecules like DNA, RNA, lipids, carbohydrates, or proteins. Nuclear Magnetic Resonance (NMR) spectroscopy has been developed to study these metabolites. NMR uses very strong magnetic fields to study the atomic changes in a metabolite. Each metabolite thus produces its own particular "NMR spectra," or fingerprint, derived from various biofluids such as urine, plasma, and saliva. NMR is used to produce "metabolic profiles" for many metabolites following toxic chemical exposures, linking them with disturbances pointing to disease.[9]

Proteomics studies the complete set of proteins that exists in a network of interactions in the cell. The cell's protein complement (proteome) is much more complex than its corresponding genome, with approximately half a million proteins being generated from approximately 30,000 human genes. Proteins constitute the majority of a cell's physical mass and have very specific three-dimensional structures that can be modified after synthesis, allowing them to perform the various tasks. Because proteins cooperatively form complexes, many protein-protein interactions occur in the cell. These functional protein networks change when a cell is exposed to toxic compounds.[10] Various high-throughput technologies, such as mass spectroscopy (MALDI-TOF)[11] and chip-related technologies, are used to study toxic perturbations in the human proteome.[12]

Bioinformatics has arisen to analyze the enormous flood of data arising from the development of high-throughput biotechnologies. Bioinformatics warehouses these huge compilations of data and mines them for their significant patterns. Our understanding of the large-scale molecular patterns of living cells then becomes possible. Researchers see this computational work as necessary to develop an "in-silico" toxicology or a way to predict the risk for adverse health effects from exposures using computer-intensive methods.[13]

ENFORCING RIGHTS IN A
NEW BIOTECHNOLOGY LANDSCAPE

As with other new science and technologies, environmentalists may soon integrate the advances of the biotechnology revolution, including the "omics" data underlying the emerging risk assessment, into their work. Hence, environmentalists will have a unique opportunity to influence the mostly unknown consequences of the new "omics" technologies. The following four recommendations are examples of how environmentalists may use Article 5 of the Genetic Bill of Rights to set the direction of their influence of "omics" consequences in this new technology landscape.

Making "omics" databases publicly accessible: Just as the human genome sequence was made public,[14] so must environmental health–relevant "omics" data also be made public. Communities must have access to "omics" data for decision-making, given their key role in the reformulation of risk assessment.

Participatory toxicity consensus: There are concerns that inaccurate toxicity claims will result from misinterpretations of "omics" data.[15] This can be addressed by developing an adequate scientific consensus on using "omics" data to classify chemicals that emerges from collective efforts of all sectors with a stake in such designations, including communities.[16]

A right to know the functional genome: The human genome, as "part of the common heritage of humanity"[17] that "underlies the fundamental unity of all members of the human family,"[18] is a vital part of the commons. The "functional genome" and the evolved mechanisms by which the genome interacts with the environment are also part of the commons. Nearly twenty years ago, legislators affirmed that "Every American has the right to know the chemicals to which they may be exposed."[19] The "right to know" may be extended by linking it to the functional genome. Given the "omics" efforts under way, greater access to the relevant databases will increase the new risk assessment's power. By expanding the existing Toxics Release Inventory's access to private hazardous materials data to include access to public and private "omics" databases, the government will have greater environmental protection power and citizens will be further empowered to act on the new information.

Genotyping: The ongoing efforts to develop individual genome sequencing or genotyping are becoming increasingly sophisticated, faster, and cheaper.[20] Given these rapid developments, individual genotyping will become increasingly common and form part of medical, nutritional, and consumer practices, while producing novel information on existing racial and ethnic populations. Eventually, mass genotyping may serve environmental health by identifying susceptible populations. This era's environmental protection might assign a population's most vulnerable genotypes as the minimal safety standard.

THE COMING TRANSFORMATION

Because chemical use is a key aspect of the global economy, its proper regulation has great importance. The foundation of this regulation in the twenty-first century is undergoing a transformation. Science is expected to provide a sound mechanistic basis for regulation, from which emerge the policies, legislation, and enforcement forming the regulatory architecture. Thus, the new genome-based biotechnologies are setting a new foundation for regulation, and the shape of that coming edifice is as yet unknown.

The uncertainty in the regulatory outcome causes concern in the various sectors. The corporate sector's concern is made clear with several key strategic questions that are the foundation of research initiatives that are described as shaping the chemical industry's future:

- How do we reduce the scientific uncertainties that can lead to overly conservative regulations?
- What scientific risk assessment processes can be developed that would more readily be understood and trusted by the public?[21]

Alternatively, civil society worries that the new biotechnology-based risk assessment could allow greater chemical deregulation.

The right of people to live without undue exposures that are genetically harmful emerges in a challenging period of history. This period is characterized by the unprecedented growth of scientific knowledge and technological prowess as well as the globalization of economic power and democratic struggles. What role will science and technology have in the evolution of democratic rights? The answer in large part depends on whether citizens and scientists can forge a path that expands and protects human rights in this post-genome era.[22]

NOTES

1. Council for Responsible Genetics (CRG), "The Genetic Bill of Rights," www.gene-watch.org/programs/bill-of-rights.html.

2. I would revise Article 5 of the Genetic Bill of Rights to make it more suitable to actual circumstances. It would read as follows: "All people have the right to live free from undue exposures to hazardous chemical or physical agents that can damage or alter the structure or function of their genetic material." This revision recognizes that humans have essentially unavoidable mutagenic exposures, such as irradiation by cosmic rays and UV light, cellular production of oxygen radicals, and the ingestion of certain plant compounds and naturally found minerals.

3. P. Montague, "Risk Assessment—Part 1: Early History of the Chemical Wars," *Rachel's Environment and Health News* 194 (1990); P. Montague, "Risk Assessment—Part 1: The Emperor's Scientific New Clothes," *Rachel's Environment and Health News* 393 (1994).

4. S. Brudnoy, "Pushing for a Paradigm Shift in Cancer Risk Assessment," *Scientist* 7 (1993): 14.

5. K. Olden and J. Guthrie, "Genomics: Implications for Toxicology," *Mutation Research* 473 (2001): 3–10.

6. M. D. Schena, D. Shalon, R. W. Davis, and P. O. Brown, "Quantitative Monitoring of Gene Expression Patterns with a Complementary DNA Microarray," *Science* 270 (1995): 467–70.

7. K. Olden, J. Guthrie, et al., "A Bold New Direction for Environmental Health Research," *American Journal of Public Health* 91 (2001): 1964–67.

8. K. S. Ramos, "EHP Toxicogenomics: A Publication Forum in the Postgenome Era," *EHP Toxicogenomics* 111 (2003): A13.

9. J. P. Shockcor and E. Holmes, "Metabonomic Applications in Toxicity Screening and Disease Diagnosis," *Current Topics in Medicinal Chemistry* 2 (2002): 35–51.

10. N. Plant, "Interaction Networks: Coordinating Responses to Xenobiotic Exposure," *Toxicology* 202 (2004): 21–32.

11. S. Kennedy, "The Role of Proteomics in Toxicology: Identification of Biomarkers of Toxicity by Protein Expression Analysis," *Biomarkers* 7 (2002): 269–90; and R. W. Nelson, D. Nedelkov, et al., "Biosensor Chip Mass Spectrometry: A Chip-Based Proteomics Approach," *Electrophoresis* 21 (2000): 1155–63.

12. J. H. Ng and L. L. Ilag, "Biomedical Applications of Protein Chips," *Journal of Cellular and Molecular Medicine* 6 (2002): 329–40; and H. Zhu and M. Snyder, "Protein Chip Technology," *Current Opinion in Chemical Biology* 7 (2003): 55–63.

13. M. D. Barratt and R. A. Rodford, "The Computational Prediction of Toxicity," *Current Opinion in Chemical Biology* 5 (2001): 383–88; and G. M. Pearl, S. Livingston-Carr, and S. K. Durham, "Integration of Computational Analysis as a Sentinel Tool in Toxicological Assessments," *Current Topics in Medicinal Chemistry* 1 (2001): 247–55.

14. A. Pollack, "Scientist Quits the Company He Led in Quest for Genome," *New York Times*, January 23, 2002, C1.

15. L. L. Smith, "Key Challenges for Toxicologists in the 21st Century," *Trends in Pharmacological Sciences* 22 (2001): 281–5.

16. R. Tennant, "The National Center for Toxicogenomics: Using New Technologies to Inform Mechanistic Toxicology," *Environmental Health Perspectives* 110 (2002): A8.

17. B. M. Knoppers, "Population Genetics and Benefit Sharing," *Community Genetics* 3 (2000): 212–4.

18. UNESCO, Human Dignity and the Human Genome, Universal Declaration on the Human Genome and Human Rights (1997).

19. U.S. Environmental Protection Agency, "Emergency Planning and Community Right-to-Know (EPCRA)," www.epa.gov/compliance/civil/federal/epcra.html; and U.S. Environmental Protection Agency, "EPCRA Overview," yosemite.epa.gov/oswer/ceppoweb.nsf/content/epcraOverview.htm.

20. E. Jonietz, "Personal Genomes: Individual Sequencing Could Be Around the Corner," *Technology Review* 104 (October 1, 2001): 30.

21. Chemical Industry Institute of Toxicology (CIIT), "Overview of CIIT," www.ciit.org/about/overview.asp.

22. For more extensive treatment of these issues, see José F. Morales, Genomic Justice, www.pibiotech.org.

Part VI

EUGENICS

"All people have the right to protection against eugenic measures such as forced sterilization or mandatory screening aimed at aborting or manipulating selected embryos or fetuses."

Article 6 of the Genetic Bill of Rights

· 15 ·

Procreative Autonomy Versus Eugenic and Economic Interests of the State

Ruth Hubbard

$\mathscr{B}$ecause of its inclusiveness, the right against eugenic measures (expressed in Article 6 of the Genetic Bill of Rights) raises far more complex issues than may be apparent at first glance. At this time in U.S. history, few liberally minded people would have problems granting individuals the right to be protected against forced sterilization. However, disagreements are likely to arise about whether "all people" is really meant to include everyone. This is especially true with reference to people judged to be of "low" intelligence or "mentally ill." To the degree that a social service or medical agency or, indeed, a court of law would certify someone to be incapable of reasoned thought, as that concept is understood by the sort of educated, middle-class person who would be called upon to assess that individual's decision-making capacities, some supporters of this right might not want to see it extended to such people. Presumably, the case for denying it would be based on the grounds that people with a handicap of this sort cannot be expected to exercise adequate foresight when engaging in sexual activities. They might be thought to be more likely than the average person to produce children when they are not able to care and provide for them adequately. The argument would continue that society needs to "protect" such people against what would constitute involuntary procreation rather than involuntary sterilization.

This line of argument implicitly assumes that, to be truly voluntary, an action must be literally willed (the term *voluntary* being derived from the Latin word *voluntas*, "will") and that people with certain mental (in)capacities are necessarily unable to foresee the potential consequences of their sexual activities sufficiently to truly "will" to have children, with full understanding of what the decision entails. Confronted with people who have deficits of this sort, society

141

might be said to have not just the right but the duty to defend them against their shortcomings by making it impossible for them to procreate—in other words, sterilizing them.

This line of reasoning is dangerous in view of the fact that laws permitting, or indeed mandating, compulsory sterilization of people deemed mentally "defective" were enacted and enforced against disempowered people in the United States, Nazi Germany, and the Scandinavian countries during much of the twentieth century.[1] This unfortunate history suggests that it is safer for democratic societies to err on the side of possibly "excessive" permissiveness by insisting on everyone's procreative autonomy.

There is a further point. All compulsory sterilization laws, though often couched in protective—albeit at times overly paternalistic—language, leave no doubt about their economic motivation. Proponents of such laws may claim that they are intended to improve the genetic endowment of future generations by preventing the birth of "hereditary defectives." Yet detailed, but necessarily hypothetical, economic projections of the financial "burden" of caring for people unable to care for themselves leave no doubt that the laws are, in no small measure, intended to protect the state from having to assume economic responsibility for children whose parents are taken to be, by definition, incapable of caring for them. This presumption itself rests on the assumption that people of "low" intelligence or who have mental "dysfunctions" do not know how to care for themselves or their children and cannot learn to do so. It ignores the fact that appropriate social and economic measures can enable people with a wide range of mental or physical (dis)abilities to support, and care for, themselves and their families.

The demand for "mandatory screening" similarly embodies both eugenic and economic considerations. Standard arguments for preventing the birth of a disabled child regularly factor in—overtly or implicitly—the expenses of caring for such children that are likely to fall, at least in part, upon state services. Yet these expenses vary widely, depending on the nature of the disability and the parents' economic and social situation. The public expense therefore cannot be estimated with any degree of accuracy.

References to the threat that "bad genes" present for society's gene pool are not always made explicit. Geneticists and pediatricians, however, tend to accept the need to reduce their presence as a legitimate social goal and to be anything but non-directive in the way they present the relevant information to prospective parents. As a result, it takes a good deal of determination on the part of prospective parents to resist the medical, family, and wider societal pressures to do all they can to avoid producing a child with a foreseeable "disability" or "defect."[2] Indeed, human geneticists have often argued that medical measures which keep infants with hereditary "defects" alive cannot truly benefit society unless and until it is possible to detect such "defects" early enough to either prevent the children's birth or to medically alter their genetic makeup so that they at least cannot transmit their own "defects" to future generations.

Given the medicalized outlook that is virtually universal in the technologically developed countries, few people who have the economic possibility to undergo genetic testing or screening (and in the future perhaps also genetic manipulation) prenatally or after birth are likely to refuse the supposed benefits of these interventions. Nonetheless, I would guess that, at present, in liberal democracies there would be considerable opposition to making the interventions mandatory. Yet here again, the economic argument, combined with the prevailing eugenic unease, could tip the balance in favor of legally mandating these interventions as they become more customary and less expensive.

A further issue is likely to end up limiting the choice of future parents to refuse such interventions. This problem will arise if egg collection and preimplantation or other prenatal interventions come to be considered standard medical practice, especially if they are deemed necessary to save the life or prevent a severe disability of the future child. Indeed, it is not clear that people could refuse such interventions for an embryo or fetus, lodged within a pregnant woman, provided the procedure does not necessitate puncturing or cutting the woman's body, as does amniocentesis or fetal surgery.

In the United States, the law usually supports a physician who insists on treating a child over the parents' objections, if there is a medical consensus that, without treatment, the child is likely to die or be irreversibly damaged. It seems entirely possible that similar arguments could be successfully extended to in vitro eggs, sperm, or embryos.

In the 1970s and 1980s, a number of women were forced to undergo court-mandated cesarean sections, against their will, when the attending physician argued that the fetus/baby would not survive a vaginal birth.[3] (It is worth noting that in none of the instances in which the woman in question managed to escape the operation by either hiding or spontaneously going into labor before the operation could be scheduled did the baby, in fact, suffer any damage.) When this practice was finally challenged in court, the Federal Appeals Court in Washington, D.C., ruled against the use of court-mandated cesareans.[4]

Perhaps pregnant women could use this decision as a precedent on which to base refusing a recommended medical test or treatment for their fetus, if the intervention would involve breaching their bodily integrity. They would, however, be unlikely to succeed if they tried to use it for embryos in vitro. It is questionable whether future parents would have the right to refuse to have the embryos biopsied, if testing embryos produced in vitro came to be considered standard medical practice. Physicians' assertion that the procedure was necessary for the health or survival of the future child could give them the same right to medical intervention they now have in the case of "at-risk" children. On what grounds could future parents base the right to refuse eugenic measures, such as screening, discarding, or genetically manipulating their in vitro embryos, for which this proviso of the Genetic Bill of Rights argues?

I think the right in question merits support because it constitutes an essential part of reproductive autonomy as well as of autonomy in matters of health, but I am not sure it would withstand a legal challenge.

Further problems are likely to arise in the context of the technical and social innovations that have redefined the status, rights, and obligations of parenthood. In the ordinary situation, U.S. jurisdictions, in general, consider the husband of a pregnant woman to be the father of her child, irrespective of tests performed to establish biological paternity. A man who wishes to contest that status must take legal steps to do so. This recognition, however, predates the use of present-day reproductive interventions. These raise the possibility that a child can have up to five potential parents: the two people who together decide to have a child as part of their family, the woman and man who provide the eggs and sperm used for in vitro fertilization, and the woman who gestates the in vitro embryo and gives birth to the child.

Confronted with a test case of this sort in which none of the five were willing to assume parental responsibilities for the born child, a California appeals court ruled that a couple who use reproductive technologies to have a child are considered to be that child's parents.[5]

Does that mean that these two persons are the only people with the right to make the relevant decisions about the fate of the embryos or the fetuses? Can they alone decide that the embryos should be biopsied and which of them are to be gestated, or whether some should be discarded? Can they insist that the surrogate mother undergo amniocentesis, or be subjected to fetal surgery or other invasive interventions, if there are medical reasons to do so? What of the possibility of fetal reduction because the legally defined parents had not planned to have more than one child? Presumably, all such eventualities need to be weighed and covered in the contracts governing these kinds of transactions, but this not-so-hypothetical situation makes it clear that the right being weighed here stands on shaky legal ground.

Going beyond this situation to decisions about the uses to which human embryos left over from attempted pregnancies, or produced commercially, can be put either in scientific research or to produce "stem cells" for possible therapeutic purposes, it is again far from clear who has the right to make these decisions, or how to enforce that right. There is much in this right to keep legal and medical specialists in reproductive technologies busy for years to come.

Finally, this right could come to constitute a major challenge to the movements supportive of women's right to make their own reproductive choices. In the United States, a woman's right to abortion, guaranteed in *Roe v. Wade* (1973),[6] establishes that right, irrespective of her reasons for the decision, as long as it is made in consultation with a physician. Surely, this right must cover reproductive decisions that may or may not lead to terminating the pregnancy.

Some of us may deplore the fact that scientific research is producing increasing numbers of eugenic grounds for that kind of decision. We may further be outraged that economic and other social pressures are pushing women who can afford to pay for their abortions toward "choice eugenics," while poor women increasingly are deprived of the possibility of obtaining an abortion, whatever their reason. Nonetheless, it seems clear that, for advocates of abortion rights, the right to terminate a pregnancy, whatever the reasons, must trump our opposition to eugenic interventions—provided the decision to accept such an intervention is "voluntary" and entered into by "choice." But, I cannot forgo putting "voluntary" and "choice" in quotation marks, because economic and other societal pressures clearly steer such decisions. Individuals, and specifically women, may feel compelled to make eugenic decisions based on circumstances that are entirely beyond their control. Nonetheless, they, and they alone, must be the ones to make them.

NOTES

1. Phillip R. Reilly, *The Surgical Solution: A History of Involuntary Sterilization in the United States* (Baltimore: Johns Hopkins University Press, 1992); and Daniel J. Kevles, *In the Name of Eugenics* (New York: Knopf, 1985).

2. Ruth Hubbard and Elijah Wald, *Exploding the Gene Myth* (Boston: Beacon Press, 1993, 1998).

3. Veronika E. B. Kolder, Janet Gallagher, and Michael T. Parsons, "Court-Ordered Obstetrical Interventions," *New England Journal of Medicine* 316 (1987): 1192–6; and Janet Gallagher, "Prenatal Invasions and Interventions: What's Wrong with Fetal Rights," *Harvard Women's Law Journal* 10 (1987): 9–58.

4. *In Re A.C., Appellant,* 573 A.2d 1235 (D.C. Court of Appeals 1990).

5. *In Re Marriage of Buzzanca,* 72 Cal. Rptr. 2d 280 (Cal. App. 4th Dist. 1998).

6. *Roe v. Wade,* 410 U.S. 113 (1973).

· *16* ·

A Disability Rights Approach to Eugenics

Gregor Wolbring

*F*orced sterilization was extensively imposed on both disabled and non-disabled people and is still currently inflicted on disabled people, especially so-called "mentally disabled people."[1] The arguments, however, have changed. Historically, forced sterilization was imposed mostly to prevent the birth of more disabled people, a eugenic argument.[2] The justification in common use today is that forced sterilization is done for the benefit of the procreating person. It is reasoned that she or he won't be able to cope with a child and it would be harmful to the child to be brought up by the disabled person. Strictly speaking, this is not a eugenic argument and would not be covered by Article 6 of the Genetic Bill of Rights issued by the Council for Responsible Genetics (CRG). It is not a eugenic argument because the focus is not the betterment of the gene pool, the race, or even the abilities of the person to be sterilized, but it is done for the quality of life of the child and for the benefit of the person to be sterilized. It could be called a quality of life argument.[3]

Mandatory screening aimed at aborting or manipulating selected embryos or fetuses is socially undesirable and unethical. Although it might be a while until we see implementation of mandatory screening for everything labeled a disease, defect, and disability, the informal pressure for such comprehensive screening is increasing as the procedures become a routine part of standard maternal care. In an opinion survey of genetic counselors and other genetic practitioners around the world, there was widespread agreement with the statement: "It is socially irresponsible knowingly to bring an infant with a serious genetic disorder into the world in an era of prenatal diagnosis."[4] Although there is no legal definition of "serious,"[5] more than 50 percent of the respondents in South Africa, Belgium, Greece, Portugal, the Czech Republic,

Hungary, Poland, Russia, Israel, Turkey, China, India, Thailand, Brazil, Colombia, Cuba, Mexico, Peru, and Venezuela agreed with the statement, as did 26 percent of U.S. geneticists, 55 percent of U.S. primary care physicians, and 44 percent of U.S. patients.[6] The above results seem to verify the assumption that informal pressure exists toward screening and termination of pregnancy or embryo selection.

There are a few problems I find with Article 6 of the Genetic Bill of Rights or, at the very least, the CRG's interpretation of that provision. It looks as if there may be an inconsistency in the CRG's adoption of Article 6 and its support of a petition that demands the prohibition of pre-implantation genetic diagnosis (PGD) for non-medical reasons. Signed by a coalition of groups, the petition states the following:

> Our organizations, which work to promote the rights, health, and reproductive freedoms of women, believe that condoning the non-medical use of PGD for sex selection would promote an already controversial technology for an inherently discriminatory use. While motivations for desiring a child of a particular sex may vary, there are no non-sexist reasons for pre-selecting sex except in cases of preventing serious sex-linked diseases. This is true even in the United States, where economic and social pressures to raise male children are minimized in comparison to other societies.[7]

By itself, this statement implies that the medical use of PGD (which includes the use against the characteristic disability) is condoned. We all know that there is no way to quantify serious disease, so that statement condones, in essence, what Article 6 tries to prevent: namely, eugenics measures. This line of argument also increases genetic discrimination, which seems to be in disaccord with Article 8 of the Genetic Bill of Rights.[8] Although people might also be in favor of the prohibition of PGD for so-called medical reasons, support of the above petition makes no statement regarding the application of PGD for such uses. In other words, no statement exists by CRG to demand the prohibition of PGD for disability deselection or ability selection. By signing the above statement, CRG is only on record so far to prohibit PGD for "social reasons" such as sex selection.

Article 6 of the Genetic Bill of Rights offers no recourse against informal pressure as well as people acting on their own prejudices, developed during their life, to use PGD and prenatal predictive testing. The quote below illustrates this prejudice nicely:

> How did parents endure the shock [of the birth of a thalidomide baby]? The few who made it through without enormous collateral damage to their lives had to summon up the same enormous reserves of courage and devotion

that are necessary to all parents of children with special needs and disabilities; then, perhaps, they needed still more courage, because of the special, peculiar horror that the sight of their children produced in even the most compassionate. Society does not reward such courage . . . because those parents' experience represents our own worst nightmare, ever since we first imagined becoming parents ourselves. The impact upon the brothers and sisters of the newborn was no less horrific. This was the defining ordeal of their family life—leaving aside for now the crushing burden on their financial resources from now on.[9]

Parts of the disability community are most assuredly ahead on that issue. One of the resolutions from the Sixth World Assembly of Disabled People International, a group representing millions of disabled people, issued in October 2002, states: "We believe that no parent has the right to design and select their unborn child to be according to their own desires and no parent has the right to design their born child according to their own desires."[10]

Similar are the demands from the Ascender Alliance, the first disabled transhumanist group,[11] and the Solihull declaration from 2000, which states:

- "We demand an end to the bio-medical elimination of diversity, to gene selection based on market forces and to the setting of norms and standards by non-disabled people."
- "Biotechnological change must not be an excuse for control or manipulation of the human condition or bio-diversity."
- "An absolute prohibition on compulsory genetic testing and the pressurising of women to eliminate—at any stage in the reproductive process—unborn children who, it is considered, may become disabled."
- "That disabled people have assistance to live—not assistance to die."
- "That having a disabled child is not a special legal consideration for abortion."
- "That no demarcation lines are drawn regarding severity or types of impairment. This creates hierarchies and leads to increased discrimination of disabled people generally."[12]

Article 6 of the CRG's Genetic Bill of Rights also falls short because it only relates to genetics and assumes eugenics is based on genetic characteristics. That by itself, of course, is not the whole story. We use ultrasound to prevent the birth of humans who might be born without legs, arms, with cleft palate,[13] and other non-genetic conditions, again based on public prejudice.

For disabled people, Article 6 of the Genetic Bill of Rights would have been more effectively stated if it were to read: "All people have the right to have been conceived, gestated, and born without being judged on their genetic and non-genetic composition." My alternative formulation of Article 6 ensures that no one who will be born would have been engineered based on the wishes of the parents or society. This formulation of Article 6 would mean the prohibition of germline genetic manipulation but also the prohibition of in utero somatic, genetic, and non-genetic intervention. It would mean the prohibition of prebirth genetic selection (as in embryo selection after pre-implantation genetic diagnostics) and prebirth genetic and non-genetic dese-lection (as in termination of pregnancy after prenatal testing).

NOTES

1. See www.bioethicsanddisability.org/sterilisation.html.

2. Dr. Margaret Thompson, Order of Canada member and former president of the ge-netics society of Canada, said as the defending witness for the Alberta government in the Leilani Muir sterilization case "some causes of mental effectiveness are hereditary and when the eugenics board was created there was a real danger of passing on those causes because contraceptive choices were limited. Today, people at risk of inheriting or passing on a defect to their children have the pill and other contraceptives available. They can seek genetic counseling before a child is born and can abort a child likely to be defec-tive." *Calgary Herald*, June 29, 1995; see also www.bioethicsanddisability.org/submissi.html and www.bioethicsanddisability.org/sterilisation.html.

3. Dorothy C. Wertz, "Eugenics Is Alive and Well: A Survey of Genetic Profession-als Around the World," *Science in Context* 11, nos. 3–4 (1998): 493–510.

4. European Commission, "The Ethical Aspects of Prenatal Diagnosis: Opinion of the Group of Advisers on the Ethical Implications of Biotechnology" (Brussels, 1996), quoted in M. Pembrey, "In the Light of Preimplantation Genetic Diagnosis: Some Eth-ical Issues in Medical Genetics Revisited," *European Journal of Human Genetics* 6 (1998): 4–11; point 67, Draft Report on Pre-implantation Genetic Diagnosis and Germ-Line Intervention, presented at the Ninth Session of the International Bioethics Committee of UNESCO (IBC) in Montreal, Canada (November 27, 2002), www.unesco.org/ibc/en/actes/s9/ibc9draftreportPGD.pdf.

5. Wertz, "Eugenics Is Alive and Well."

6. See www.genetics-and-society.org/resources/cgs/2002_asrm_sex_selection.html.

7. European Commission, "The Ethical Aspects of Prenatal Diagnosis."

8. Although the sign on to the petition by itself might not indicate that the CRG condones the so-called medical usage of PGD, I think it is fair to conclude that the CRG seems to make a distinction between medical and non-medical use, as no state-ment exists from the CRG where they demand the prohibition of the medical use of PGD.

9. T. Stephens and R. Brynner, *Dark Remedy: The Impact of Thalidomide* (Cambridge, MA: Perseus, 2001): 65–6.

10. Text available on request from Disabled People International, at dpi@dpi.org.

11. See groups.yahoo.com/group/Ascender_Alliance/files/ASCALLI-07JAN02-MANIFESTO-2.doc and groups.yahoo.com/group/Ascender_Alliance/files/ASCALLI-06SEP02-ANTIEUGENICS-RACISM%20POLICY.doc.

12. See www.johnnypops.demon.co.uk/bioethicsdeclaration/index.htm.

13. C. Stoll, C. A. Wiesel, A. Queisser-Luft, U. Froster, S. Bianca, and M. Clementi, "Evaluation of the Prenatal Diagnosis of Limb Reduction Deficiencies," *EUROSCAN Study Group Prenatal Diagnosis* 20 (2000): 811–18; 70.8 percent of limb reduction deficiencies were terminated, and 100 percent of the Phocomelia were terminated, www.ncbi.nlm.nih.gov:80/entrez/query.fcgi?cmd=Retrieve&db=PubMed&list_uids=11038459&dopt=Abstract; Z. Blumenfeld, I. Blumenfeld, and M. Bronshtein, "The Early Prenatal Diagnosis of Cleft Lip and the Decision-Making Process," *Cleft Palate Craniofacial Journal* 36, no. 2 (1999): 105–7.

Part VII

GENETIC PRIVACY

"All people have the right to genetic privacy including the right to prevent the taking or storing of bodily samples for genetic information without their voluntary informed consent."

Article 7 of the Genetic Bill of Rights

17

Genetic Privacy in the Health Care System

Jeroo Kotval

The advent of genetic testing has been accompanied by concerns about unauthorized dissemination of this information and especially about use of this information to discriminate in areas such as insurance and employment. In the realm of health care, these anxieties have been heightened by the twin moves to create a computerized medical record and a market-driven health care system. This chapter revisits the function that confidentiality has traditionally played in the health care system and explores why genetic information warrants concern; it then briefly examines the institutional context in which the genetic information exists, and why this threatens the traditional conception of genetic privacy. The discussion is limited to concerns regarding the health care system, not only because it is here that the information is most likely to be generated and first used, but because the issues elaborated in this context can be readily applied to other institutional uses and misuses of genetic information.

PRIVACY

Privacy in health care has been vigorously defended but inadequately justified. This is partly because privacy is a complex concept with no universally accepted definition. But it is also because in a complex, liberal, democratic society, privacy is taken for granted, as perhaps it should be, were it not for threats to it from advances in both genomic technologies and computerization along with the rise of market-driven health care.

Anita Allen has delineated the concept of privacy in the health care setting into four distinct dimensions:[1]

- Physical privacy relates to the notion of seclusion, solitude, and freedom from unwanted and unwarranted contact with other people.
- Informational privacy is expressed in practices that impose limits on the accessibility of personal information revealed to another and what we think of as confidentiality.
- Decisional privacy means that individuals ought to be able to make certain personal decisions free from intrusions or coercion by third parties.
- Proprietary privacy asserts the individual's right with respect to their stored biological samples and the information obtained from these.

Most of the discussion here is restricted to concerns related to informational privacy, since this is likely to be of the most immediate concern to the largest number of people. The pervasiveness of privacy practices in the health care setting—from gestures such as drawing curtains during examinations and prohibiting unauthorized persons from being present at the bedside, to codifying privacy in medical ethics and crafting rules for maintaining medical record confidentiality—emphasizes how central privacy concerns are in health care. These practices serve to create an atmosphere of trust, comfort, respect, and care.

Privacy protects our interests in our individuality and in our social, economic, and political relationships. All conceptions of privacy imply a domain of the personal as contrasted with the public sphere, include the concept of limited access by others to one's body or mind, and imply freedom from unwarranted and unauthorized intrusion. Even though definitions of privacy are value neutral, privacy is most often seen as beneficial in the health care setting. All definitions of privacy acknowledge that it is a value that is necessary to individuality and the proper development of our personality. Jeffrey Reiman makes this claim when he says that "privacy is a social ritual by means of which an individual's moral title to his existence is conferred. Privacy is an essential part of the complex social practice by means of which the social group recognizes—and communicates to the individual—that his existence is his own."[2]

CONFIDENTIALITY

Even though we loosely talk about confidentiality and privacy as if they were interchangeable concepts, they are often distinguished in the literature. In the health care setting, confidentiality is one form of privacy—that is, informational privacy—and through hundreds of years has been a central concept in informing our practices and policies about professional integrity, effective care, and public health and safety.

Confidentiality may be conceptualized as the communication of private personal information from one person to another when it is expected that the recipient of the information, such as a health professional, will not ordinarily disclose the information to a third person.[3] The distinction between privacy and confidentiality is that one must relinquish privacy by revealing information to someone before confidentiality comes into play. Confidentiality creates a comfort zone in which privacy may be surrendered within a limited sphere to achieve a positive outcome and implies that there is a relationship based on trust between at least two persons. It ensures that the patient has control over his or her privacy by allowing the patient to determine what information should be revealed, to whom, when, and how.[4] Privacy, therefore, affords a measure of freedom and autonomy to the patient by providing discretion over how he or she manages human relationships. Since a professional person ought to do what it takes to nurture the professional-client relationship, it follows that confidentiality is justified because patients' interests demand it.

The above discussion is not meant to suggest that modern-day practices are consistent with the theoretical explorations of privacy in health care. In fact, the foregoing discussion is intended to draw out departures of the ideal from practice. Privacy in health care is always in balance with other competing interests that serve the common good. These include public health interests as well as those that provide assurance of quality care and fiscal prudence. In the modern-day health care system, however, privacy is also in the balance with competing interests that may have a debatable relevance to quality health care, such as profits for investors in market-driven, profit-driven health care institutions.[5]

WHY WE SHOULD CARE

Genetic information is going to be increasingly part of the medical record because of expanding uses of genetic tests. The benefits of testing are several and include greater control of reproductive choices, by virtue of greater diagnostic and prognostic power. The major source of genetic information about an individual is likely to result from an analysis of DNA or RNA in the process of seeking health care, as a research subject, upon involvement with the law, or from an analysis of biological specimens stored in DNA banks. The more traditional avenue of genetic information, such as family histories, will continue to play a role, however, as genomic science advances.

Without getting caught up in debates about whether genetic information is unique or exceptional compared to other medical information, we can say at the very least that:

- Genetic information is not only uniquely personal, but it also reveals personal information about an individual's close family members.
- Some genetic information—especially results of those tests that give advance information about future health conditions—has tremendous social power,[6] in that it has the capacity to stigmatize and lends itself to discrimination, especially in the areas of employment and insurance. And in a system such as ours, where most persons obtain their health insurance through their employment, some types of genetic information could render some persons both unemployed and uninsured.
- The social power of genetic information gains force from the pervasive belief in our culture that our genes are us—that they reveal something deep, basic, and final about us. Genetic information is sensitive information because our relationship to our biological selves is culturally conditioned and fraught with personal meaning—and our very attitudes toward genetic information are culturally shaped. Genetic information is sensitive information also because it can serve to stigmatize a racial group—albeit very inaccurately, but the list of genetic conditions that are represented disproportionately in certain racial groups, creating a crude racial road map, continues to grow. Given the history of perverse social uses of genetic information, this warrants concern. While education may serve to dispel some of this power, its wellsprings are deeper than can be dispelled through cognitive measures. In short, violation of genetic privacy may have a different impact on persons, depending on their race, ethnicity, and social class.
- Privacy of the medical record is a crucial value because, if it is not assured, people will hesitate to obtain genetic tests and so will not avail themselves of the benefits of such tests.

Two points, however, must be acknowledged. First, not all genetic information has the same level of social power. For example, genetic information predicting an individual's likelihood of developing leukoderma, a skin discoloration, does not have the same power as information predicting a severely crippling, high-cost illness such as Huntington's disease, Alzheimer's disease, or certain mental illnesses. Second, certain non-genetic medical information also has great social power, such as a record of a sexually transmitted disease (STD) or a visit to an emergency room with a gunshot wound or drug overdose.

Nevertheless, genetic information differs from non-genetic information in that genetic information has a long life. This is so not just because DNA is a stable molecule; rather it is because genetic information is not something one recovers from and puts behind one—unlike most STDs or a gunshot wound. Genetic information continues to implicate one's immediate relatives even after one is dead and gone—especially when it is part of an institutional electronic medical record, which can be cross-linked to others who may be implicated. The life of an institutionally maintained record may span several generations.

It is not feasible to protect genetic information separate and apart from the rest of the medical record: (1) because distinctions between genetic and non-genetic bases of a health condition are increasingly blurred, (2) because the contours around a definition of genetic information cannot be adequately drawn without making such a definition over-inclusive or under-inclusive of what can appropriately be called genetic information, and (3) most importantly, because genetic information cannot be sequestered apart from the rest of the medical record without sacrificing the benefits of genetic testing. The solution is to safeguard the confidentiality of the entire medical record.

CONCERNS OF MARKET-DRIVEN HEALTH CARE

It is important to examine the systemic context in which medical information exists, since how a technology is used is often determined by the imperatives facing the institution using this technology. The dramatic shift in enrollment from not-for-profit to for-profit health plans in an environment of market-driven health care demands that we take a closer look at how advance information about future health care costs may be used in this institutional environment.

The ethos of cost savings in health care is a laudable goal when it comes to eliminating practices that do not provide a return in improved health and are therefore wasteful. Investor-sponsored, for-profit health care institutions, however, must turn a profit to secure their financial well-being. The very structure of this institution therefore lends itself to the suspicion that organizational decisions related to health care may not be made solely with the interest of the patient at heart.

Genetic testing can provide information before the advent of any symptoms. While much of this information is probabilistic, and likely to remain so for some time, it does suggest a risk. Those intent on cost savings may not find the absence of certainty a deterrent to barring access to the benefits of insurance at an affordable price. The Health Insurance Portability and Accountability Act of 1996 and the privacy regulations ensuing from it provide some protection to persons enrolled in group health insurance plans from a denial of health insurance based on genetic information. Persons seeking insurance in private markets can still suffer exorbitant premium costs or be denied access unless their state laws provide protections. An in-depth analysis of the benefits and limitations of the law is, however, beyond the scope of this chapter.

The departure from the ideal of confidentiality is obvious. As described earlier in this chapter, the intent of health care confidentiality is to ensure that patients do not fear harm to themselves from revealing their secrets in the process of obtaining health care. Confidentiality is intended to secure trust so that persons may feel free to reveal their secrets and so that the provider can make an accurate assessment before providing care. As I have discussed at length previously, trust in the physician is now a matter that is complicated by trust in the health care institution, which has the power to create a culture of cost savings with all that this implies.[7]

NOTES

This work has been supported by the U.S. Department of Energy, Office of Health and Environmental Research, Grant Number DE-FG02-97ER62430 to Jeroo S. Kotval.

1. Anita Allen, "Genetic Privacy: Emerging Concepts and Values," in Mark A. Rothstein, ed., *Genetic Secrets* (New Haven: Yale University Press, 1997).

2. Jeffrey H. Reiman, "Privacy, Intimacy, and Personhood," *Philosophy and Public Affairs* 6, no. 1 (1976): 26–44.

3. William J. Winslade, "Confidentiality," in Warren T. Reich, ed., *Encyclopedia of Bioethics* (New York: Simon and Schuster Macmillan, 1995), 452.

4. Alan F. Westin, *Privacy and Freedom* (New York: Atheneum, 1967). Westin's often-cited definition of privacy is that privacy is the claim of individuals, groups, or institutions to determine for themselves when, how, and to what extent information about themselves is communicated to others (p. 7).

5. Jeroo S. Kotval, "Market-Driven Managed Care and Confidentiality of Genetic Tests: The Institution as Double Agent," *Albany Law Journal of Science and Technology* 9, no. 1 (1998): 1–25.

6. Dorothy Nelkin, "The Social Power of Genetic Information," in Daniel J. Kevles and Leroy Hood, eds., *The Code of Codes: Scientific and Social Issues in the Human Genome Project* (Cambridge, MA: Harvard University Press, 1992).

7. Kotval, "Market-Driven Managed Care and Confidentiality of Genetic Tests."

18

Biotechnology's Challenge to Individual Privacy

Philip Bereano

$\mathcal{A}$lthough the legal concept of a "right to privacy" was originally conceived by Louis Brandeis as merely the "right to be left alone," in the intervening century since he put that idea forward it has been substantially elaborated to encompass more philosophical elements, such as human integrity, dignity, and autonomy. All four of these aspects are essential to the articulation of a "right to genetic privacy." The rise of modern biotechnology as a fundamentally contextualizing technology—significantly challenging older notions of the human relationship to nature, and infusing all aspects of modern culture, such as language, movies, literature, and legal concepts of property—has led to significant challenges to individual privacy.

Several paradigms can be used to explain the social nature of technologies. Our understanding of these will help in an analysis of the human impacts of the new genetics. Until the end of World War II, technology was equated with "progress" and thus was seen as a good in itself. People didn't look for much more complicated relationships. U.S. culture celebrated technological "advances" without hesitation. Then, as the result of increasing and increasingly recognized "externalities" and the critical perspectives provided by analysts such as Ralph Nader and Rachel Carson, who articulated some of the technological dysfunctionalities people began to experience in their daily lives, this old paradigm collapsed.

Some people advocated a radical "greening" or "anti-technological" point of view, that we have "enough" technology or should renounce new technical possibilities. Despite the idealistic appeal such visions had for some people, these paradigms had limited circulation, since they did not correspond well to social reality.

In place of the old widespread ideology of progress, liberal commentators developed the currently dominant view, the "use-abuse" model: technologies in themselves are neutral and value free but they can be used for the good or misused to the detriment of society. While technological change is generally undertaken for beneficial ends, we may need social regulation to ensure that it is not misused. This is the unarticulated view underpinning TV shows such as *Nova*, science reporting in major newspapers, and motivating professional organizations such as the American Association for the Advancement of Science and the policy writings of its journal *Science*.

Of course, some of the negative impacts of technologies experienced by some individuals (e.g., genetic testing techniques resulting in uninsurable people) are intentionally developed by other social actors. This realization—that we are not all in this together—has led to the development of a "social relations" paradigm of technology and society. In this view, technologies are developed (through research and development programs, public funding, specially favorable laws, etc.) because they reflect the needs and goals of powerful segments of society. These agendas include, of course, social control of those people with less power.

In addition, as many in the Council for Responsible Genetics (CRG) have shown, biotechnology is based on a specific ideology of genetic determinism, namely the view that configurations of genes determine one's future and define who one really is. When this view is coupled with the rapaciousness of current capitalism and globalization (as in the strikingly radical notion of patenting genes, which gives power and control over actual human genome segments to corporations) and with a testing mania sweeping society (which is an undisguised manifestation of power relationships), it should not be surprising that major forces are working to obtain genetic information from the vast majority of us—to sort and categorize us, manipulate us, stratify us, reward or punish us, and weed out the "damaged goods" in order to eugenically "improve the race"—all in violation of our privacy. This is why the articulation of a right to genetic privacy is currently so compelling. It is clear that modern biotechnology reflects the social relations paradigm. It is about control—over nature, the human species, and unruly individuals.

Despite the passage of the Americans with Disabilities Act, disabled people have been put on the defensive by a series of Supreme Court setbacks and the growth of a eugenics movement based on new "liberal" biology. The grandiose claims of biologists that miracle cures will surely follow if they are allowed to do cloning and stem cell research reemphasize a view of women as vehicles for men's profit-making activities. New laws are being passed to expand the pool of persons who must involuntarily submit to the taking of DNA samples for entry into databanks. So genetics is increasingly becoming a component of technocratic class politics.

In the 1990s, when the U.S. Department of Justice began prodding the states to establish DNA ID databanks—under the Combined DNA Index System (CODIS)—the rationale used to popularize them was the protection of women against rapists, even though the FBI had no track record that showed a particular concern for women, and despite the reality that in the majority of rapes the identity of the perpetrator is known (a relative or boyfriend). But the databanks were never, in fact, limited to sexual assaults. As the CRG anticipated, the technocratic logic inherent in these systems inevitably extended the scope of the databanks to encompass other felonies, then all convicts, and now there are calls to include all arrestees (a measure already enacted in several states). The CRG worked with the public defender in Massachusetts to oppose such a bank but lost in the courts.[1] The latest proposal for anti-terrorism legislation from the Bush administration (called the Patriot II Act) would involve the taking of DNA from virtually everyone in the country.[2]

The basic rationale for these databanks is to reduce future crime. But the historic U.S. public policy recognizes that other values limit our deployment of unfettered policing of the population. For example, the Fourth Amendment protects against unreasonable searches and seizures, and the Fifth Amendment protects against self-incrimination. It would seem that only databanks of criminals convicted of crimes (1) with a high recidivism rate and (2) likely to leave samples should be rationally justified. Thus, the expansion of databanks shows that DNA privacy is becoming non-existent before the demands of social control and the ideology of genetic determinism.

Since 1992, the Department of Defense (DOD) has maintained a DNA ID bank based on the fiction that such a registry would ensure that there would never be another "unknown soldier." An unarticulated justification was actually to assemble corresponding body parts of persons killed in military actions and gory mishaps and attacks for burial. From the beginning, however, the bank has covered civilian employees as well, and the data originally were to be held for seventy-five years, a time span that was reluctantly reduced to fifty years after complaints. The CRG worked with a number of service personnel who desired to protect their genetic privacy and not be included in the databank—including the first two resisters.[3] All were unsuccessful except for those who could make a plausible freedom of religion argument, including one sailor who practiced Native American rites and was harassed for years until the DOD quietly dropped the matter as publicity mounted. Apparently, Jehovah's Witnesses and Christian Scientists are automatically excused from the program. Despite pressure from the CRG and others, the DOD has refused to issue regulations regarding protections for the data in the files—for example, that there would be no release to third parties.

The invasion of genetic privacy is not an isolated phenomenon. Other technologies of surveillance and control are increasingly employed by the powerful interests against the rest of us—TV cameras filming public places, wiretaps (without judicial warrants), and devices for monitoring Internet communications and surfing. People are being conceptualized as packets of information—objects to be surveilled, categorized, manipulated, and punished for any variation from the norms laid down by societal powers.

Clearly, we are no longer being "left alone." The forcible taking of biological samples violates people's integrity. Our manipulation as objects of control violates our dignity. Our increasing powerlessness violates our autonomy. The ideology of technological perfectionism is riding high right now.

As the late Tony Mazzocchi, one of the CRG's founders, explained twenty years ago: "The problem with any [genetic] screening and surveillance program is that it depends on who controls it and who administers it. In a perfect world genetic screening might be a very adequate [occupational health and safety] surveillance measure—it could be used to protect people. However, this is not a perfect world."[4] As if corroborating Mazzocchi's prescience, the recent cases *Norman-Bloodsaw v. Lawrence Berkeley Laboratories*[5] and *EEOC v. Burlington Northern and Santa Fe Railway Co.*[6] indicate employers were using genetic screening without worker knowledge in order to exercise inappropriate control over the labor force. One can argue that these actions are not merely "abuses" of the technology (as the dominant corporate liberal paradigm would suggest) but are inherent in its nature as a technique for employers manifesting their power, consistent with the social relations model.

The American Civil Liberties Union has recently developed a new Databases and Civil Liberties policy that sums up these privacy concerns and is applicable to genetic information:

> The systematic collection of personal information—which, through the use of powerful modern computer systems can be easily searched, maintained in great quantities, rapidly cross-correlated, and aggregated or disaggregated—has become a troubling phenomenon with many adverse implications for civil liberties. This systematic collection, storage, and dissemination of personally sensitive information can greatly enhance the power of governments, corporations, institutions and individuals to infringe rights such as privacy, autonomy, equality, and fairness in criminal and other proceedings. . . .
>
> . . . The ability of an individual to exercise control over the collection, maintenance, and use by the government of his or her sensitive personal in-

formation is central to personal integrity and human dignity. Personal autonomy is threatened when government or the private sector embraces modern technologies that facilitate the large-scale collection, correlation, duplication, and use of personal information.

. . . As a general principle, personal information should not be collected from individuals without their informed consent. Mandatory collection of personal information must be limited to the minimum required to achieve legitimate public policy objectives. Exceptions to the informed consent principle may occur in the furtherance of an important legitimate public policy objective, but should only be allowed with specific statutory authority. Information collected under these circumstances must be destroyed or rendered anonymous as soon as the legitimate and proximate reason for its collection is satisfied.

Whenever possible, government agencies or their surrogates or corporations that collect personal information should do so directly from the individual, rather than a third party, and should use the least intrusive method possible.

Thomas Jefferson warned us that "vigilance is the eternal price of liberty." As new technologies have been developed, the forms of vigilance we require have had to evolve as well. Privacy has become a heightened region of contested terrain in an increasingly "technologized" world. Our genes are the latest privacy battleground.

NOTES

1. *Landry v. Attorney General*, 429 Mass. 336 (Supreme Judicial Court of Massachusetts, 1999).

2. Domestic Security Enhancement Act of 2003 (Draft Legislation).

3. *Mayfield v. Dalton*, 109 F.3d 1423 (9th Circuit, 1997).

4. Quoted in Terri Goldberg, "Genetic Power to the People: An Interview with Tony Mazzocchi," *GeneWatch* 15, no. 1 (2003): 15.

5. *Norman-Bloodsaw v. Lawrence Berkeley Laboratory*, 135 F.3d 1260 (9th Circuit, 1998).

6. *EEOC v. Burlington Northern and Santa Fe Railway Co.*, No. 01-4013 (N. Iowa filed Feb. 9, 2001).

Part VIII

GENETIC DISCRIMINATION

"All people have the right to be free from genetic discrimination."

Article 8 of the Genetic Bill of Rights

· *19* ·

Beyond Genetic
Anti-discrimination Legislation

Joseph S. Alper

$\mathcal{S}$ometime in the mid-1980s, a group of scientists and other academics (including me) interested in the social implications of science formed the Genetic Screening Study Group. We realized that the almost unbelievably rapid progress in molecular genetics would inevitably lead to technological applications that could have adverse effects on large groups of people. In particular, we became concerned with the potential for what we called "genetic discrimination."

GENETIC DISCRIMINATION IS
A REAL AND SERIOUS PROBLEM

We defined genetic discrimination as "discrimination on the basis of genotype." In formulating this definition, we wished to distinguish genetic discrimination from discrimination that people experience as a result of being, or being perceived to be, unhealthy. People experiencing genetic discrimination are currently healthy; they show no symptoms of the illness associated with altered genes they may carry.

In order to establish whether genetic discrimination existed, and if so, whether it was a real problem, we began with a pilot survey and then conducted an extensive survey of approximately 27,000 people who either had genetic diagnoses or were professionals such as medical geneticists and genetic counselors. Our questionnaire and follow-up telephone interview were designed to elicit information about incidents of genetic discrimination.

Our results were published in two papers, Billings et al. (1992)[1] and Geller et al. (1996).[2] These papers were the first to document the existence of genetic discrimination. We found cases in which people were denied life and health insurance, were not hired for jobs, and were denied admission to schools. A particularly clear and repeated example of genetic discrimination involved the denial of health and life insurance to people with hemochromatosis, a recessive genetic condition, which when untreated results in excess iron stored in the body. When treated by phlebotomy (blood drawing), people with hemochromatosis can live normal and healthy lives. The morbidity (illness) and mortality rates of people with treated hemochromatosis are no higher than the rates of the general population.

We concluded from our surveys that genetic discrimination is a real and growing problem, especially in the area of health insurance. We foresaw that tests for genetic susceptibilities to common diseases such as cancer and heart disease would be developed, and we predicted that insurance companies would be tempted to use the results of these tests to deny health insurance to those with positive test results. In addition, we were concerned that people who would benefit from a genetic test that detects the presence of an altered gene for a serious but treatable disease before symptoms appear might decide not to submit to a genetic test for fear of losing their insurance. To summarize the problem in its starkest form, presently healthy people with positive genetic tests may not be able to obtain or afford insurance, and people whose health care might benefit from genetic test information may be too fearful of such genetic discrimination to be tested.

In our papers, we noted that the only real solution to the problem of genetic discrimination is universal health insurance coverage. However, given the dim prospects of universal health care in the United States, we also suggested that legislation be enacted banning genetic discrimination. This call was taken up by others, and currently a majority of the states have legislation banning genetic discrimination in one form or another.

In recent years, several members of the Genetic Screening Study Group, in particular Jon Beckwith and I, have been reconsidering the desirability of singling out genotype as a category of illegitimate discrimination. We have come to the conclusion that a focus on genetic discrimination is both scientifically mistaken and politically problematic. In our view, a focus on genetics as a category of discrimination has the potential of exacerbating, rather than alleviating, the problems caused by the misuse of genetic information. More detailed versions of our arguments are given in Alper and Beckwith (1998)[3] and Beckwith and Alper (1998).[4]

GENETIC ANTI-DISCRIMINATION LEGISLATION IS FLAWED

Genetic anti-discrimination legislation presupposes that genetic information, tests, and diseases are clearly distinguishable from non-genetic information, tests, and diseases. In all three of these categories, the distinction between genetic and non-genetic is difficult to maintain. In their legislation, various states have defined genetic information in different ways. Genetic information can be restricted to information about the actual genetic material itself, such as DNA and RNA. Or it can be extended to include proteins that are not genes themselves but whose synthesis is directed by genes. In some states, the definition of genetic information has even been extended to include information based on family history.

Similarly, the distinction between genetic and non-genetic medical tests is not clear-cut. Some clinical tests that are not considered genetic tests may in fact serve as indicators of an altered gene. A positive test for high cholesterol or for occult fecal blood may be the consequence of mutations in genes conferring susceptibility to heart disease or colon cancer. Some might argue that essentially all medical tests are genetic because they are designed to detect biochemicals whose synthesis has been directed either directly or indirectly by genes.

Difficulties in classifying medical tests suggest that the difference between genetic and non-genetic diseases is also not clear-cut. It is evident that cystic fibrosis and Huntington's disease are genetic diseases. In nearly all cases, an individual will develop a genetic disease if and only if the disease genotype is present. However, for multifactorial diseases such as cancers and heart conditions, genes may influence the origin and course of the disease, but no single gene can be either a necessary or sufficient cause of the disease. The interaction between the environment and such genes is still poorly understood.

We conclude that in view of these ambiguities in the distinction between genetic and non-genetic information, tests, and diseases, the privileging of genetic information in anti-discrimination legislation becomes difficult to justify. Even if the genetic/non-genetic distinction could be maintained, serious questions arise about the fairness of legislation prohibiting discrimination based on genetic information but not discrimination based on non-genetic medical information. Imagine someone who by means of a genetic test is found to carry a gene conferring susceptibility to colon cancer but is not presently affected by the disease. Under a genetic anti-discrimination law, this information could not be used to deny insurance to that individual. Now consider another person who has a non-genetic clinical test indicating a precancerous condition (which is not genetically inherited) that sometimes leads to colon cancer. For example, the condition might have been caused by environmental factors such as diet. This person would not be protected by the genetic anti-discrimination legislation. Both individuals may be at equal risk for developing colon cancer. Is it fair that one should be protected by anti-discrimination legislation simply because the diagnostic test was genetic rather than non-genetic?

GENETIC ESSENTIALISM

It seems to us that genetic anti-discrimination laws are at least partially motivated by the erroneous belief that genes play the fundamental role in determining who we are. This belief has been termed "genetic essentialism."[5] If it is true that genes play the defining role in determining our essential nature, then it is easy to argue that genetic information about ourselves should be accorded more privacy protection than any other kind of information. Such private information, it is argued, should not be used for discriminatory purposes. In addition, genetic essentialism implies that genetic characteristics, unlike non-genetic characteristics, are innate and unalterable. Proponents of genetic anti-discrimination legislation argue that people should not be blamed for their innate and unalterable characteristics. This argument seems to carry with it the invidious implication that we *are* responsible for our non-genetic diseases.

With the completion of the sequencing of the human genome, it has now become clear to almost everyone that genetic essentialist ideas are untenable. Almost daily, geneticists are finding complexities in the interactions between genes and the environment in the development of human traits and diseases. With the possible exception of the relatively rare single gene diseases, it is now evident that there are no "genes for" a particular disease. We have added the qualifier "possible" because even supposedly simple single gene diseases display remarkable complexities. The same altered cystic fibrosis gene can result in a wide variety of symptoms. Some people with the altered gene for Huntington's disease do not develop the disease.

REDRAFTING GENETIC ANTI-DISCRIMINATION LEGISLATION

We recommend that genetic anti-discrimination legislation be rewritten to prohibit discrimination based on any predictive medical information. Such redrafted legislation would serve two purposes. First, it would overcome the weaknesses present in existing genetic anti-discrimination legislation. Second, it would provide an important first step toward the goal of universal health care.

In thinking about genetic anti-discrimination legislation, we are presented with a dilemma. On the one hand, genetic anti-discrimination laws are needed because genetic discrimination is a real problem that seems certain to worsen as new genetic tests become available. On the other hand, most genetic anti-discrimination laws are flawed because (1) they are based on misconceptions concerning the distinction between genetic and non-genetic information, tests, and diseases, (2) they rely on the genetic essentialist belief that our genes define our essential being, and (3) they are unfair.

We believe that genetic anti-discrimination legislation serves an important function because it attempts to address a deeper and more pervasive underlying problem than discrimination arising from the use of genetic information alone. The problem is not merely discrimination based on genetics but rather discrimination based on any type of predictive medical information. Historically, the insurance system in the United States has always discriminated against applicants for insurance policies who already show symptoms of a condition, irrespective of whether their illness has anything to do with genes. But now, with genetic and other predictive medical tests, it is becoming possible to label asymptomatic people as diseased even though they have no symptoms of illness and may never develop the illness.

Genetic anti-discrimination legislation implicitly recognizes the importance of prohibiting discrimination based on all types of predictive medical information as opposed to prohibiting only that discrimination based on a person's genotype. As we have already noted, definitions of genetic information in some legislation have included information about family history as well as information about the genetic material itself. As a result of these expanded definitions of genetic information, some genetic anti-discrimination legislation already prohibits discrimination against the asymptomatic ill that results from the use of many types of predictive information that do not rely on molecular genetic technology or even on the existence of genes.

The expanded definitions of genetic information in the laws suggest that genetic anti-discrimination legislation is not primarily driven by worries about the use of genetics per se. Instead, people are becoming increasingly concerned about the crisis in the health care system. They are worried about the loss of medical insurance in an era when the cost of treating a disease can be astronomical. They are also worried about the privacy of their medical records when computers make it ever easier to transmit information. If the scope of the types of information prohibited by anti-discrimination laws is expanded to include all types of predictive medical information, then some significant fraction of these fears may be alleviated.

Viewed in this light, laws controlling genetic discrimination represent an effort in the direction of comprehensive reform of the health care system.

Unfortunately, in their present form, most of these laws appear to be vague, ambiguous, and unfair. In order to avoid these difficulties, which arise from misconceptions about genetics, it will be necessary to redraft many of these laws.

Redrafted legislation that prohibits labeling currently healthy people as diseased on the basis of any type of predictive medical information would accomplish the purpose of eliminating much of the discrimination against the "asymptomatic ill" that the public so much fears. Such legislation will not, of course, benefit those individuals who are already ill and cannot obtain health insurance. Nevertheless, the recognition that genetic information is not special and that problems arising from the use of genetic information cannot be separated from problems arising from the use of all other types of medical information will mark a major step toward the goal of universal health care.

NOTES

1. Paul R. Billings et al., "Discrimination as a Consequence of Genetic Testing," *American Journal of Human Genetics* 50 (1992): 476–82.

2. Lisa Geller et al., "Individual, Family, and Societal Dimensions of Genetic Discrimination: A Case Study Analysis," *Science and Engineering Ethics* 2 (1996): 71–88.

3. Joseph S. Alper and Jonathan Beckwith, "Distinguishing Genetic from Nongenetic Medical Tests: Some Implications for Antidiscrimination Legislation," *Science and Engineering Ethics* 4 (1998): 141–50.

4. Jonathan Beckwith and Joseph S. Alper, "Reconsidering Genetic Antidiscrimination Legislation," *Journal of Law, Medicine and Ethics* 26 (1998): 205–10.

5. Dorothy Nelkin and Susan Lindee, *The DNA Mystique: The Gene as a Cultural Icon* (New York: W. H. Freeman, 1995).

20

Analyzing Genetic Discrimination in the Workplace

Paul Steven Miller

In this twenty-first century, humankind wields increasing power to understand, alter, and control the world in which we live. The mysteries of our genetic code are being unveiled, providing remarkable new insights into our unique human characteristics. Rapid developments in information technology bring ready access to a vast reservoir of new data. The information age has taken hold and the genetic revolution is in full swing. With apologies to Aldous Huxley, we stand at the precipice of a *brave new world*.

It has been only fifty years since Watson and Crick's groundbreaking discovery of the double helix. The profound developments that have taken place during their lifetimes in the science of genetics are staggering. More staggering still are the potential benefits, the boundless horizons, the promised and as-yet-unimagined applications of their work, and the work of the many dedicated scientists involved in the sequencing of the human genome. There can be no doubt in my mind that a firm and unwavering commitment to the betterment of humankind has fueled this tireless effort.

DNA technology has rapidly achieved unquestioned currency in matters such as guilt, innocence, and paternity. Its further application to myriad social and legal contexts, including the workplace, presents some compelling ethical questions. As with other technological advances, there is always a risk of misapplication. History has taught us that the wonders of science hold unique power to sway and seduce and, all too often, corrupt the course of human nature. James Watson has indicated that progress in genetics should come only with a firm awareness of the potential for abuse.

As commissioner of the U.S. Equal Employment Opportunity Commission (EEOC), I was interested in the protection of those in the workplace who could be marginalized by this "progress." My work was to ensure that the integrity, the individuality, and the very being of every person is never threatened in the name of science. As we chart a new course, we must do so with an abiding respect and admiration for our potential and the rights of individuals. In short, we must insist that our genetic profile—in whatever form it takes—remain ours, and that it should never be used to violate our fundamental rights as people.

As the science of genetics explodes, and the technology becomes more accessible, the issue of how society protects its workers against the misuse of genetic information will become more important, and legal and policy development in the area will be more compelling. If employers are permitted to consider genetic information in making personnel decisions, people may be unfairly barred or removed from employment for reasons that are wholly unrelated to their ability to perform their jobs.

Experts predict that by the end of the decade, the modest sum of $100 will buy a test that effectively identifies genetic markers for a myriad of conditions and diseases. While there is some anecdotal evidence of employers making decisions based on asymptomatic genetic predispositions, there is no solid empirical evidence measuring the extent to which, or for what purposes, employers presently seek genetic information about their employees or applicants. Yet the potential for genetic discrimination is real and not just the stuff of science fiction. Employers can learn an employee's genetic information through genetic testing, company medical exams, family medical history information, or medical records. Moreover, the mere fear of discrimination may make people hesitate before accessing a growing array of genetic tests that can identify vulnerability to specific diseases.

A cornerstone of American workplace anti-discrimination law is the principle that applicants and employees be selected and evaluated on their ability and desire to do the job, and that selection not be prejudiced by myths, fears, and stereotypes based on factors such as race, ethnicity, gender, age, or disability. The Americans with Disabilities Act (ADA) prohibits discrimination against qualified individuals with a disability. The ADA does not address the specific issue of genetic discrimination. That said, and recent Supreme Court jurisprudence notwithstanding, it is axiomatic that the ADA covers individuals with a manifested genetically related impairment, assuming it substantially limits a major life activity. Similarly, the ADA arguably covers individuals with a record of a genetically related disability, such as someone who has recovered from cancer. The more challenging question is whether the ADA prohibits discrimination based on a diagnosed, but asymptomatic, genetic condition that does not substantially limit a major life activity.

In 1995, albeit in policy guidance that lacks the force of law, the EEOC adopted the view that the ADA prohibits discrimination against workers based on their genetic makeup. EEOC policy explicitly states that discrimination on the basis of genetic information is covered under the third prong of the statutory definition of the term *disability*, which addresses individuals who are *regarded as* having substantially limiting impairments. Designed to protect against prejudices and misconceptions about disability, this part of the statute reflects Congress's recognition that the reactions of others to an impairment, or a perceived impairment, may be just as disabling as an actual impairment.

Notwithstanding the foregoing argument, some worry that courts will find that the ADA does not cover genetic predisposition discrimination, or that genetic discrimination is so different from traditional disability discrimination that the ADA does not provide a satisfactory framework. However, the principle that genetic discrimination in employment be forbidden enjoys wide bipartisan support. Legislation has been introduced in the U.S. Congress by Republican as well as Democratic members that specifically prohibits discrimination on the basis of genetic information by employers. Moreover, President George W. Bush has expressed his support for genetic discrimination legislation. On October 14, 2003, by a 95 to 0 vote, the Senate passed the Genetic Information Nondiscrimination Act, which prohibits employers and health insurers from discriminating based on an individual's genetic profile.[1] The bills on the Hill are based, at least in part, upon an executive order signed by President Clinton at the end of his term that prohibits the federal government from considering genetic information in hiring, promotion, discharge, and all other employment decisions.[2] Finally, a patchwork of state laws, though they vary widely in form and breadth, offers some additional protection.

Given the inexorable progress of science, the question of whether genetic testing in the employment context is or will ever be appropriate eludes easy answers. Absent new legislation, the ADA permits disability-related inquiries and medical examinations of employees, which include genetic tests, when they are job related and consistent with business necessity. This means that an employer may only obtain medical information about its employees when it has a reasonable belief based on objective evidence that (1) an employee will be unable to perform the essential functions of his or her job because of a medical condition, or (2) the employee will pose a direct threat because of a medical condition.

The historical antecedent for the job-related standard was employers' use of information about the physical or mental condition of employees to exclude or otherwise discriminate against individuals with disabilities, despite their ability to do the job. The job-related standard gives the employer the burden to demonstrate that it is relevant and appropriate to test for a particular genetic marker. In the context of an asymptomatic genetic disorder, I believe this would be a difficult burden to meet.

Underlying the debate are fundamental questions about the proper role of the employer in employee health issues, and whether and how much our society, which has long championed the rights of individuals, should permit paternalism in the workplace. Over a decade ago, the Supreme Court provided some insight into this question in a case that addressed an employer's obligations vis-à-vis the rights of female employees of child-bearing age. In *International Union v. Johnson Controls* (1991), the Court held that a chemical company's policy barring women with the ability to bear children from opportunities that included the potential for harmful lead exposure violated Title VII as gender discrimination.[3] Dismissing concerns of potential tort liability, the Court opined that the employer's presumptive goodwill did not excuse a patently discriminatory practice.

In an ADA case, *Chevron v. Echazabal* (2002), our current Supreme Court showed increasing tolerance for workplace paternalism.[4] The case involved an individual who, for over twenty years, worked as a contractor at a California oil refinery. Chevron twice refused to hire him permanently, concluding that his liver impairment, ultimately diagnosed as Hepatitis C, might by exacerbated by exposure to the solvents and chemicals present in the refinery.

Relying on the legislative history of the ADA and the *Johnson Controls* decision, the Ninth Circuit had reversed the District Court's decision for the employer and found that Chevron was motivated by improper paternalistic concerns for Echazabal's health, and that the right to resume personal risks in the workplace resided solely with the individual.[5]

A unanimous Supreme Court disagreed. The Court upheld the employer's reliance on the "direct threat to self" affirmative defense, rejecting the Ninth Circuit's holding that the ADA precludes employers from considering the "direct threat to self."[6] In so doing, the Court found ample room for distinction between workplace paternalism and specific documented risks to employees. Quoting from the EEOC's direct threat regulation, the Court emphasized that an employer must provide an individualized analysis, based on reasonable medical judgment, that assesses an employee's current ability to perform the job and takes into account the imminence of the risk and severity of the potential harm.

No genetic employment discrimination case has ever been decided, either in U.S. federal or state court. Two years ago, the EEOC and the Burlington Northern and Santa Fe Railroad settled the first lawsuit alleging genetic employment discrimination under the ADA.[7] In that case, the EEOC alleged that Burlington Northern subjected its employees to surreptitious genetic testing to identify a genetic marker for carpal tunnel syndrome to address a high incidence of repetitive stress injuries among its employees. The EEOC further alleged that at least one employee was threatened with discipline and possible termination for refusing to submit to the test. The case settled before judgment, so the court never ruled on the applicability of the ADA to the circumstances of the workers' complaints. Particularly reassuring about the Burlington Northern case was that, amid the fury of publicity generated by the case, no one suggested that genetic testing of employees should be allowed.

In sum, while genetics may be suitable for determining paternity and guilt or innocence, genetic information should not be used to exclude qualified workers from the workplace. Genotype is no substitute for qualifications, and no employer should ever review your genetic records along with your résumé.

NOTES

1. United States Congress, Senate, S. 1053, Genetic Information Nondiscrimination Act of 2003.

2. Executive Order 13145 to Prohibit Discrimination in Federal Employment Based on Genetic Information.

3. *International Union v. Johnson Controls*, 499 U.S. 187 (1991).

4. *Chevron v. Echazabal*, 536 U.S. 73 (2002).

5. *Echazabal v. Chevron*, 226 F.3d 1063 (9th Cir. 2000).

6. *Chevron v. Echazabal*, 536 U.S. at 78–87.

7. *EEOC v. Burlington Northern and Santa Fe Railway Co.*, No. 01-4013 (N. Iowa filed Feb. 9, 2001).

21

Disability Rights and Genetic Discrimination

Gregor Wolbring

"*A*ll people have the right to be free from genetic discrimination"—that is the statement of Article 8 of the Genetic Bill of Rights of the Council for Responsible Genetics, and it reflects a global concern that the advancement in the ability of genetic manipulation and prediction leads to genetic discrimination. If people have the right to not be discriminated against based on race, gender, and other characteristics, it makes sense to also demand the right not to be discriminated against based on one's inherent genetic composition. Many countries are developing genetic anti-discrimination laws, and many international documents, such as the UNESCO Universal Declaration on Human Rights and the Human Genome, demand the prohibition of genetic discrimination. However, from a disability rights perspective, three questions arise: What is the meaning of the term "genetic discrimination"? What is covered by genetic anti-discrimination laws? Who is covered by genetic anti-discrimination laws?

THE MEANING OF GENETIC DISCRIMINATION

For me, the definition is simple: genetic discrimination occurs if we deal with humans or potential humans in a discriminatory fashion based on the knowledge, perception, or reality attached to the consequences of having a particular gene, gene activity, or gene product.[1] Discrimination is not associated with and directed against a cell, a zygote, an embryo, a fetus, a person, or a born human being[2] but associated with and directed against a *characteristic* of a cell, a zygote, an embryo, a fetus, or a born human being. As it will soon become clear, my simple definition is not what the laws in preparation and those in existence have in mind.

WHO IS COVERED BY GENETIC DISCRIMINATION LAWS?

Genetic anti-discrimination laws could prohibit discrimination directed against a *characteristic* of a cell, a zygote, an embryo, a fetus, or a born human being. These laws could prohibit discrimination against born human beings who have the gene for a disease, like Huntington Chorea, whether or not they exhibit the clinical symptoms for the condition. And these laws could prohibit discrimination against people who have the gene for the condition until such time that the clinical symptoms for the condition appear. In this case, it would be legal to discriminate against someone with the gene for Huntington Chorea once they exhibit the clinical symptoms for Huntington Chorea. In other words, as long as one is viewed as asymptomatic, one is protected from discrimination. As soon as one is viewed as symptomatic, the discrimination is not prohibited anymore.

My definition does not make a distinction between someone who gets discriminated against in the symptomatic or asymptomatic stages, and it does not make a difference as to when the genetic discrimination takes place. Genetic discrimination is wrong, period. But the law, various proposals, and the debates seem to link the concept of genetic discrimination and its unlawfulness only to the asymptomatic stage, thereby setting the stage for a very limited and narrow interpretation of the concept of genetic discrimination being the least useful for disabled people.[3] These differences in the view of acceptable discrimination based on lack of normative abilities (clinical symptoms) versus other characteristics are readily acknowledged and, it seems, endorsed by the American Civil Liberties Union (ACLU), which says in its campaign for a genetic anti-discrimination law: "Congress should take immediate steps to protect genetic privacy . . . [because] it is inherently unfair to discriminate against someone based on immutable characteristics that do not limit their abilities."[4]

Of course, disabled people in the United States and in other places do not agree with the fact that only asymptomatic people are covered by genetic anti-discrimination laws.[5] And the American College of Medical Genetics seems to follow the same sentiment as the National Council on Disability (USA).[6] These new laws as they are proposed further marginalize symptomatic disabled people from the human rights movement and further entrench a discriminatory approach of action, which is solely justified by a medical view (symptoms) of disability, something denounced, of course, by many disabled people.[7]

WHAT IS GENETIC DISCRIMINATION?

Genetic anti-discrimination laws are mostly designed to prevent genetic discrimination against asymptomatic people in regards to employment and insurance issues.[8] That might be enough for the asymptomatic; however, genetic discrimination takes place in more areas besides employment and insurance. The most obvious one is the discriminatory use of predictive prebirth genetic tests to detect a genetic variation labeled as a genetic disability—where the term *disability* is used interchangeably with terms such as *disease* and *defect*—with the following push to eliminate this genetic variation through genetic fixing or weeding-out technologies, such as selecting embryos with the desired genetic composition for IVF procedures (pre-implantation genetic diagnostic) or the termination of a pregnancy if the fetus shows an undesirable genetic composition (prenatal predictive testing). The extension of the genetic discrimination concept to the prebirth stage seems to be acceptable to justify the prohibition of the selection of the characteristic male or female (sex selection) but not acceptable for the prohibition of deselection of characteristics labeled as disabilities, defects, and diseases.[9] On the one hand, we increase the application and usage of prebirth predictive tests for "medical purposes" for the characteristics labeled as disabilities, diseases, or defects. And we try to prohibit the use of the same technologies for "non-medical purposes" such as sex selection. In other words, the society believes that some characteristics deserve special protection from the misuse of the technologies (sex selection), but others don't (disability, disease, and defect deselection).

Genetic and related discriminations also occur in wrongful birth and life suits[10] and in the acceptance of genetic therapy and enhancement (whereby enhancement could be seen as the setting of new genetic norms) for "medical reasons."[11] It seems that the prohibition of genetic discrimination, as presently conceived, is interpreted too narrowly and excludes disabled people and the characteristics labeled as disabilities, diseases, and defects. Not only does it not help disabled people, but it even does damage to them, as it is another area where they (the disabled) are pitted against the rest of the population (the non-disabled).

NOTES

1. For example, the developmental differences of the fetus exposed to the drug thalidomide leading to stillbirth or people being born with non-mainstream body compositions are thought to be caused by thalidomide changing the activity of genes and gene products during fetal development.

2. I distinguish the terms "person" and "born human being" because, from time to time, there have been attempts to link the term "person" to certain capabilities such as cognitive abilities. For example, some bioethicists would not characterize a newborn as a person because, in their mind, the newborn lacks essential capabilities. By introducing the term "born human being," I am acknowledging, albeit not justifying, these alternative bioethical definitions of "persons." For more background on the debate around personhood, see www.bioethicsanddisability.org/Personhood.html.

3. M. S. Yesley, "Protecting Genetic Difference," *Berkeley Technology Law Journal* 13 (1998): 662; and see Trudo Lemmens, "Selective Justice, Genetic Discrimination, and Insurance: Should We Single Out Genes in Our Laws?" *McGill Law Journal* 45 (2000), available at www.journal.law.mcgill.ca/arts/452lemme.pdf.

4. "ACLU Renews Calls for Congress to Ban Genetic Discrimination," February 13, 2002, available at www.aclu.org/WorkplaceRights/WorkplaceRights.cfm?ID= 9688&c=180.

5. National Council on Disability, Position Paper on Genetic Discrimination Legislation (2002), available at www.ncd.gov/newsroom/publications/geneticdiscrimination_ positionpaper.html.

6. Michael S. Watson and Carol L. Greene, "Points to Consider in Preventing Unfair Discrimination Based on Genetic Disease Risk: A Position Statement of the American College of Medical Genetics," *Genetics in Medicine* 3, no. 6 (2001): 436–7, available at www.acmg.net/resources/policies/pol-025.pdf.

7. See International Centre for Bioethics Culture and Disability, www .bioethicsanddisability.org/disabilityperception.html, and www.bioethicsanddisability .org/dislawstatistic.html.

8. Comments to the national conference of insurance legislators on the proposed genetic discrimination act, Sophia Kolehmainen, Peter Shorett, Sara Gambin, and Paul Billings, San Francisco, July 28, 2002, www.gene-atch.org/programs/privacy/ insurance.html; see also Council for Responsible Genetics, www.gene-watch.org/ programs/privacy/summary2001.html.

9. James Grifo, quoted by Gina Kolata in "Fertility Ethics Authority Approves Sex Selection," *New York Times*, September 28, 2001, accessed January 26, 2003, at www .genetics-and-society.org/resources/items/20010928_nytimes_kolata.html; G. Wolbring, "Disability Rights Approach Towards Bioethics," *Journal of Disability Policy Studies* 14, no. 3 (2003): 154–80; G. Wolbring, "Science and the Disadvantaged" (2000), www.edmonds-institute.org/wolbring.html; "Eugenics, Euthanics, Euphenics" (1999), a shorter version appeared in *GeneWatch* 12, no. 3 (June 1999); www.bioethicsand disability.org/Eugenics,%20Euthanics,%20Euphenics.html; Adrienne Asch and Eric Parens, "The Disability Rights Critique of Prenatal Genetic Testing," *Hastings Center Report*, September/October 1999; R. Mallik, *A Less Valued Life: Population Policy and Sex Selection in India Center for Health and Gender Equity* (2002), accessed January 26, 2003, at www.genderhealth.org/pubs/MallikSexSelectionIndiaOct2002.pdf.

10. G. Wolbring, expert opinion for the Study Commission on the Law and Ethics of Modern Medicine of the German Bundestag, with the title "Folgen der Anwendung genetischer Diagnostik fuer behinderte Menschen" (Consequences of the application of genetic diagnostics for disabled people) (2001), www.bundestag.de/gremien/medi/

medi_gut_wol.pdf; and "Disability Rights Approach to Genetic Discrimination," in Judit Sandor, ed., *Society and Genetic Information Codes and Laws in the Genetic Era* (Central European University Press), www.bioethicsanddisability.org/wrongfulbirth.html.

11. International Bioethics Committee of UNESCO, Report of the IBC on Preimplantation Genetic Diagnosis and Germ-Line Intervention, SHS-EST/02/CIB-9/2 (Rev. 3), April 24, 2003 (Paris: United Nations Educational, Scientific, and Cultural Organization), p. 9, paragraph 68, available at http://unesdoc.unesco.org/images/0013/001302/130248e.pdf; and "Disability Rights Approach to Genetic Discrimination," in Sandor, *Society and Genetic Information Codes and Laws in the Genetic Era*; G. Wolbring, "Science and Technology and the Triple D (Disease, Disability, Defect)," in *Converging Technologies for Improving Human Performance: Nanotechnology, Biotechnology, Information Technology and Cognitive Science* (Netherlands: Kluwer, 2003), 206–15; www.wtec.org/ConvergingTechnologies/.

Part IX

EXCULPATORY DNA EVIDENCE

"All people have the right to DNA tests to defend themselves in criminal proceedings."

Article 9 of the Genetic Bill of Rights

22

A Fundamental Right to Post-conviction DNA Testing

Peter J. Neufeld and Sarah Tofte

It is far better that a few innocent men remain in prison or are executed than that we open the judicial floodgates to inmates claiming they were wrongfully convicted.

—A high-ranking assistant district attorney
offering advice to congressional staff

Since its introduction into U.S. courtrooms in the late 1980s, forensic DNA testing has profoundly enhanced the truth-seeking function of the criminal justice system. The Federal Bureau of Investigation reports that 25 percent of samples sent to the Bureau's DNA lab for testing result in a suspect's exclusion.[1] Even higher exclusion rates are reported by other law enforcement agencies and by private labs retained by police and prosecutors. In the last decade alone, thousands of factually innocent persons would have been convicted, some sentenced to death and many others to lengthy prison sentences, but for the intervention of DNA tests. Although a DNA profile alone is not necessarily conclusive, context is critical: in sexual assault and sexual assault homicide cases, the nature of the biological material (spermatozoa, blood, or other bodily fluids or cells) and the location from which it was recovered can produce overwhelming evidence of guilt or innocence.

Just as pretrial DNA testing has illuminated the unexpected frequency with which police and prosecutors target the wrong person, post-conviction testing in cases that were tried before the advent of forensic DNA can exonerate the wrongly convicted, possibly identify the real perpetrator, and shed light on the causes of wrongful conviction.[2] As of this writing, 140 imprisoned men and one woman have been cleared through post-conviction testing.[3] Collectively, they had served more than 1,550 years in prison; thirteen of them, having been sentenced to death, had been awaiting execution when DNA serendipitously intervened on their behalf.[4] Not only had their trials produced wrongful convictions, but the state and federal appellate courts that reviewed these convictions affirmed them, often commenting on the overwhelming evidence of guilt.

These exonerations are merely the proverbial tip of the iceberg, because in 75 percent of our Innocence Project cases, we find ourselves helpless to do anything for the inmate because the critical evidence has been lost or destroyed. When evidence does still exist, in half our cases police refuse to help us locate it and prosecutors oppose testing it.[5] Their efforts to deny defendants the opportunity and ability to pursue freedom from wrongful convictions are unconscionable, given the incontrovertible evidence that innocent people have been and continue to be arrested, tried, and convicted. These attempts offer us the opportunity to view the perverse priorities practiced by some public officials. Preserving a conviction is more important to them than seeking the truth. However abhorrent their position, it is, nevertheless, at least from their perspective, understandable. For every time we uncover a wrongful conviction, the victim is again assaulted, the community roiled. The pus and bile of systemic failure are exposed: an innocent man's freedom (or life) was stolen; the real perpetrator was not caught, remains at large, and perhaps continues to prey on the community; and the manner of misconduct and negligence by police, prosecutors, crime lab technicians, or assigned defense counsel is exposed as contributory to the miscarriage of justice.

Our adversaries are resourceful and popular; they traditionally present themselves as Men in the White Hats, and they oppose post-conviction testing with the zeal of politicians in cover-up mode. When prosecutors refuse to cooperate, we have turned to the legislatures and the courts to ensure that every inmate has the right to access and test evidence that has the power to prove actual innocence.

STATE AND FEDERAL STATUTES

A general policy and principle of our criminal justice system is to make it exceedingly difficult to gain access to potentially exculpatory evidence after a conviction, particularly if the evidence is in the hands of the police. Strictures upon scant judicial resources, coupled with a popular presumption that the jury got it right the first time, have produced numerous statutory and court-created obstacles to reopening a case, if even for the very limited purpose of gaining access to biological evidence for testing.[6]

Some state legislatures, however, have seized upon the unprecedented power of DNA testing to carve out a statutory exception to the general principle. The first states to adopt such legislation were New York in 1994,[7] and Illinois in 1995.[8] In 1999, a number of states followed suit and, as of January 2004, thirty-five states had post-conviction DNA testing statutes.[9] The provisions vary from state to state, but typically a statute allows testing in the following circumstances: DNA evidence relating to the crime exists and is in a condition that allows for testing; such evidence was not previously subjected to DNA testing, or was subjected to a more primitive form of testing than that being requested; the evidence can be tracked through a clear chain of custody from crime scene to preservation to testing laboratory; there is a reasonable probability that the defendant would not have been convicted if DNA testing had been available before trial; and the request was not made for the purpose of delaying an execution.[10]

None of the states grants testing as an absolute right. These testing laws, however, vary widely in scope and substance. Some are comprehensive, state-funded testing programs open to all convicted persons with reasonable claims of innocence. Yet in others, the right to DNA testing is sharply limited: for example, by leaving decisions about testing to the sole discretion of prosecutors, or restricting testing only to a limited class of cases or under oppressive time limits. Most statutes unfairly deny access to defendants who pled guilty even if the plea was to avoid the risk of capital punishment. Because the original convictions in 35 of the 140 DNA exonerations relied, in part, on false confessions and several actually pled guilty,[11] denial of access will result in many innocent people remaining in prison or being executed. Some states have restrictive interpretations of the requirement that the evidence, if tested, could produce a reasonable probability that the defendant would not have been convicted.[12] Courts in such states ignore the fact that a DNA exclusion trumps most other kinds of inculpatory evidence. Thus, if there were a confession, or multiple eyewitnesses, or if a prosecutor were to invent a new theory of guilt that contradicts even the testimony of the victim at trial, some courts in these states would find the inmate's application wanting.

Other state statutes were enacted with "sunset provisions," legislating such a narrow window of time in which to prepare and file DNA testing petitions that few convicted persons will be able to take advantage of them.[13] Indeed, three of these statutes have already expired, just a year after their enactment, with no more than a handful of petitions filed in each state.[14]

This patchwork response from the states has lent additional momentum to a national effort to have the right to DNA testing codified by Congress. On October 1, 2003, the Innocence Protection Act (IPA) of 2003 was introduced in Congress as part of a larger bill called the Advancing Justice Through DNA Technology Act.[15] This groundbreaking bill would grant any inmate convicted of a federal crime the right to petition a federal court for

DNA testing to support a claim of innocence. The IPA would require all states who receive a variety of federal criminal justice funding to, among other things, enact post-conviction DNA testing laws that meet or exceed the bill's own standards for federal prisoners. Since the IPA's standards are quite comprehensive (including both DNA testing and preservation-of-biological-evidence requirements), its passage would require significant revisions in most of the state DNA statutes currently on the books. In November 2003, the Advancing Justice Through DNA Technology Act (HR 3214) passed the House of Representatives with an overwhelming 357–67 vote. As of this writing, it is stalled in the Senate, despite its strong bipartisan support.[16]

THE COURTS: A CONSTITUTIONAL RIGHT
TO POST-CONVICTION DNA TESTING

Under a traditional legal approach, an inmate could seek a new trial if there were newly discovered evidence of innocence or through a writ of habeas corpus in state or federal court. But the time in which one may achieve this relief is extremely tight. In Virginia, for instance, it is a mere twenty-one days after conviction. In most states, if the new evidence is discovered more than three years after the jury verdict, the inmate is time barred.

In theory, if an inmate can meet neither requirement, a grant of executive clemency from the governor provides a safety valve against the worst miscarriages of justice. But governors are political creatures, loath to release anyone from prison, fearful of being "Dukakis-ed," branded by an opponent as soft on crime. The policy reasons for narrowing the window of post-conviction relief are derived from the "doctrine of finality."[17] Under this doctrine, after appeals have been exhausted, the case is deemed over. Victims and the state are afforded closure. With the passage of years, witnesses disappear and memories fade.

In the pre-DNA era, when the fact finder relied primarily on testimony, it would be no more likely, were the verdict vacated and the case remanded for a second trial, that a second verdict would be accurate. But DNA testing eliminates most, if not all, of the doctrine of finality's concerns. Unlike witness testimony, DNA does not become weaker with the passage of time. Indeed, the dozens of DNA exonerations demonstrate that, a decade or more after conviction, DNA results are more reliable than eyewitnesses, confessions, and questionable forensic science introduced at the original trial. Crime victims read about the many exonerations and wonder about the integrity of their verdicts. A DNA test can confirm the correctness of the verdict or can trigger the release of an innocent person and put the police on the track of the actual perpetrator.

DNA, in limited situations, offers the criminal adjudicatory process a doctrine of certainty to replace the doctrine of finality. This possibility of achieving certainty demands that we push the legal envelope to establish a right to post-conviction testing. Given the moral and legal right of all innocent persons to remain free from confinement or state-ordered execution, the courts must explicitly recognize such a right by guaranteeing all defendants a procedure enabling them to pursue innocence claims through post-conviction DNA testing.

Apart from any legal right, there is a universal right for an innocent person to remain free from jail when he has not committed the crime for which he was convicted. This is a concept so basic to our notions of fairness, liberty, and justice that it hardly bears mentioning except to those law enforcement officials who continue to prevent defendants from pursuing claims of innocence. As a legal matter, the Eighth Amendment to the Constitution prohibits cruel and unusual punishment. The amendment is violated if the state is allowed to execute or keep in prison innocent men and women.

The due process clauses of the Fifth and Fourteenth Amendments require "that government action, whether through one agency or another, shall be consistent with the fundamental principles of liberty and justice which lie at the base of all our civil and political institutions and not infrequently are designated as 'law of the land.'"[18] Constitutional absolution for the concealment or destruction of exculpatory evidence is unconscionable. It fails to protect the innocent, fails to assist the apprehension of the guilty, and fails to safeguard the judicial process as committed ultimately to the ascertainment of truth.

Post-conviction DNA testing has not only led, as of this moment, to the exoneration of 140 prisoners in the United States but to the identification of a number of the real perpetrators, many of whom were serial murderers and rapists.[19] It is simply good law enforcement. In matters where analysis of biological samples that could only have come from rapists is at issue, DNA testing can definitively cut to the truth, obviating the need to make decisions by assessing the credibility of witnesses, or the imperfect recollections of traumatized victims.

A district attorney's obstruction of this effort is especially disturbing because it is fundamentally inconsistent with the prosecutor's mandate: to seek truth, apprehend the guilty, and prevent the wrongful imprisonment of the innocent.

The Supreme Court's landmark decision in *Brady v. Maryland*[20] established that suppression by the prosecution of evidence favorable to an accused, especially when specifically requested, violates due process when the evidence is material either to guilt or punishment. Under a traditional *Brady* analysis, a defendant must prove three conditions to establish a due process violation: (1) the prosecution withheld or suppressed evidence; (2) the evidence is favorable; and (3) the evidence is material to the defense. One and three are easily met. The only wrinkle with DNA is that, absent testing, there is no way to ascertain whether the evidence is "favorable."

But prosecutors are not free to ignore their due process obligations simply because, at the time they deny access, the exculpatory potential of evidence has yet to be realized. The Supreme Court has made clear that police violate due process when they, in bad faith, fail to "preserve evidentiary material of which no more can be said than that it could have been subjected to tests, the results of which might have exonerated the defendant."[21] Larry Youngblood, who was convicted of rape, challenged the destruction of rectal swabs that were collected from the victim in that case. There was never any dispute that Youngblood had a constitutional right to conduct scientific testing on the swabs had they been preserved, because the swabs constituted potentially exculpatory evidence. Indeed, under *Youngblood*, blocking access to existing evidence whose potential exculpatory value is self-evident is no different than, in bad faith, destroying it.

Even with the establishment of a constitutional right to post-conviction DNA testing, inmates nevertheless need a vehicle to have their petition heard in court. Since the conventional writs of habeas corpus will be time barred, the vehicle we have chosen is a federal civil rights statute, Section 1983,[22] passed at the conclusion of the Civil War for the express purpose of giving blacks access to federal courts to vindicate their civil rights.

If a prisoner should challenge the fact or duration of his confinement or seek relief, which would practicably invalidate his conviction or automatically lead to his immediate or speedy release, the courts could construe a Section 1983 petition as a writ of habeas corpus and dismiss it as time barred rather than allowing it to proceed as a Section 1983 action. But since a Section 1983 action only seeks access to evidence, a ruling for the inmate does not automatically lead to setting aside the conviction and effecting his release. It only provides him with the opportunity to test. Since the test results could, and in approximately half the post-conviction cases do, confirm guilt, the inmate's Section 1983 action should not be deemed a habeas petition, and thus he should evade dismissal. We have just begun to raise these issues in the courts with modest success.

Forensic DNA testing has revolutionized the way we determine guilt or innocence. The growing number of pre- and post-trial DNA exonerations proves the very real danger of convicting the innocent. Case studies of the exonerations spotlight the systemic causes of wrongful convictions. But the adverse reactions of some prosecutors to an inmate's request for access to testing also illuminate a way of thinking that promotes the preservation of convictions over the freedom of the innocent. Preservation of convictions at any cost is unconscionable and repugnant to principles of fairness, justice, and truth. At the expense of the factually innocent, a blind bow to non-binding legal precedent, factual loopholes, and outdated doctrine can no longer be accepted. We can do better.

POSTSCRIPT

After this chapter was written, Congress passed and President Bush signed into law the Justice for All Act. H.R. 5107 includes the major components of the Innocence Protection Act, another piece of congressional legislation, which, among other things, grants any inmate convicted of a federal crime the right to petition a federal court for DNA testing to support a claim of innocence. It also encourages states with the power of the purse to adopt adequate measures to preserve evidence and make post-conviction DNA testing available to inmates seeking to prove their innocence.

Other key provisions include helping the states in the forty death-penalty jurisdictions across the country to create effective systems for the appointment and performance of qualified counsel, together with better training and monitoring for both the defense and the prosecution. It provides substantial funding to the states for increased reliance on DNA testing in new criminal investigations and increases the amount of compensation available to wrongly convicted federal prisoners. Importantly, the bill also requires states seeking funding under many of its provisions to certify the existence of governmental entities capable of conducting independent external investigations of state and local crime labs where there are serious allegations of misconduct or gross negligence.

NOTES

1. United States Department of Justice, National Institute of Justice, *Convicted by Juries, Exonerated by Science: Case Studies in the Use of DNA Evidence to Establish Innocence After Trial*, Pub. No. 161258 (June 1996), xxvii; confirmed by phone conversation with Paul Bresson, FBI public information officer, January 22, 2004.

2. See generally H. Patrick Furman, "Wrongful Convictions and the Accuracy of the Criminal Justice System," *Colorado Lawyer* 32 (2003): 11, 12 (identifying causes of wrongful convictions as mistaken identification, ineffective representation, police and prosecutorial misconduct, perjured testimony, corruption of scientific evidence).

3. See www.innocenceproject.org.

4. See www.innocenceproject.org.

5. See Stephen F. Smith, "Cultural Change and 'Catholic Lawyers,'" *Ave Maria Law Review* 1 (2003): 31, 48 (prosecutors concerned about reelection and win-loss records often oppose post-conviction relief); Milton Hirsch, "Small-Town Florida 1963: Time It Was and What a Time It Was," *Champion* 27 (2003): 42 (prosecutors and law enforcement opposed to statute that provides for DNA testing, claiming potential chaos from amount of petitions they would receive); Jennifer L. Weiers and Marc R. Shapiro, "The Innocence Protection Act: A Revised Proposal for Capital Punishment Reform," *NYU Journal of Legislation and Public Policy* 6 (2003): 615, 622 (prosecutors typically oppose DNA testing for death-row inmates).

6. See generally Margaret Berger, "Lessons from DNA: Restriking the Balance Between Finality and Justice," in David Lazer, ed., *DNA and the Criminal Justice System: The Technology of Justice* (Cambridge, MA: MIT Press, 2004); Deanna F. Lamb, "Timely Justice: The Balance Between Claims of Actual Innocence and Finality of Judgments," *Lincoln Law Review* 28 (2000–2001): 17 (highlighting recent decisions that favor diligence and equity while perhaps sacrificing accuracy in judgments).

7. N.Y. Crim. Proc. § 440.30 (McKinney 1994).

8. 725 Ill. Comp. Stat. Ann. 5/116 (1995).

9. www.innocenceproject.org/docs/IP_Legislation_Memorandum.html.

10. www.innocenceproject.org/docs/IP_Legislation_Memorandum.html.

11. www.innocenceproject.org.

12. See, for example, N.Y. Crim. Proc. § 440.30 (1994) (requires "a reasonable probability that the verdict would have been more favorable to the defendant"); Ariz. Rev. Stat. § 13-4240 (2000) ("A reasonable probability exists that the petitioner would not have been prosecuted or convicted if exculpatory results had been obtained through deoxyribonucleic acid testing"); N.J. Stat. Ann. § 2A:84A-32a (must raise a "reasonable probability that the verdict or sentence would have been more favorable if the results had been available at the time of trial"). Many statutes are even more restrictive. For example, Kentucky's statute only applies to those petitioners who were sentenced to death; Ky. Rev. Stat. Ann. § 422.285 (2002). Ohio's statute requires that the DNA testing be outcome determinative, in that "no reasonable factfinder would have found the inmate guilty of that offense"; Ohio Rev. Code Ann. § 2953.71 (2003).

13. See, for example, Ohio Rev. Code Ann. § 2953.73 (2003) (Ohio allows for DNA testing applications for a period of one year, to expire October 29, 2004); Mich. Comp. Laws § 770.16 (2001) (Michigan deadline for DNA applications is January 1, 2006); 2001 Or. Laws 697 (Oregon deadline is January 1, 2006).

14. Idaho Code § 19-4902(b) (2001). The statute of limitations for Idaho applications is July 1, 2002, or one year from day of judgment, whichever is later. Fla. Stat. Ch. 925.11 (2001). The original statute of limitations in Florida was October 1, 2003; an emergency order of the Florida Supreme Court in late September 2003 extended the stay. Currently the final ruling is pending, and bills have been introduced to extend/eliminate the deadline. N.M. Stat. Ann. § 31-1A-1(2001). District courts in New Mexico would not accept any petitions for DNA testing after July 1, 2002. See also Cragg Hines, "There Should Be No Deadline for Justice," *Houston Chronicle*, September 30, 2003, available at www.chron.com/cs/CDA/printstory.hts/editorial/2129849

(criticizing Florida's two-year window for post-conviction DNA testing by describing the story of Frank Lee Smith, a wrongly convicted man who died of cancer while on death row); and "Issue a Stay of Injustice," *Palm Beach Post*, September 24, 2003, 10A ("An innocent man doesn't become guilty just because the clock hits midnight on a certain date").

15. Advancing Justice Through DNA Technology Act of 2003, H.R. 3214, 108th Cong. § 3 (2003).

16. See "Justice Through DNA," *Washington Post*, October 7, 2003, A24.

17. See Berger, "Lessons from DNA: Restriking the Balance Between Finality and Justice"; and Lamb, "Timely Justice: The Balance Between Claims of Actual Innocence and Finality of Judgments." See also Anne-Marie Moyes, note, "Assessing the Risk of Executing the Innocent: A Case for Allowing Access to Physical Evidence for Posthumous DNA Testing," *Vanderbilt Law Review* 55, no. 3 (2002): 953, 995 (criticizing the doctrine of finality as limiting a defendant's right to challenge his conviction).

18. *Hebert v. Louisiana*, 272 U.S. 312, 316–17 (1926).

19. In at least thirty-six cases across the country, the exoneration of an innocent person led to the discovery of the true perpetrator. In the case of Ronald Cotton, when DNA from the victim's vaginal swab and underwear was entered into a national database of known felons, a man who had previously confessed to the same crime came up as a match; www.innocenceproject.org/case/display_profile.php?id=6. New York City's Central Park Jogger case is another example of DNA confirming the true perpetrator's confession to the crime. When Matias Reyes, already in prison for rape and murder, confessed to the rape of the Central Park jogger, his DNA was determined to be a match to semen found on the victim's sock. Testing of hairs found at the scene also matched Reyes. "'It's Time': New Confession Casts Doubt on Convictions in 1989 Central Park Attack," *abcNEWS.com*, September 26, 2002.

20. *Brady v. Maryland*, 373 U.S. 83 (1963).

21. *Arizona v. Youngblood*, 488 U.S. 51, 57 (1988).

22. 42 U.S.C. § 1983.

23

Forensic DNA: The Criminal Defendant's Right to an Independent Expert

John Tuhey

Carl Jung once said of science that it "is not . . . a perfect instrument, but it is a superb and invaluable tool that works harm only when taken as an end itself."[1] The science of forensic DNA is not a perfect instrument, but it is often viewed as one. One can easily speculate the reasons for this phenomenon. Perhaps it is because of DNA's ability to do what many earlier forensic methods (e.g., blood or fiber testing) were unable to do in the past, which is to take minute samples of evidence and narrow down the field of matching suspects to a finite few.[2] Evidence of blood or semen found at a crime scene can become very compelling to a juror when an expert explains that the likelihood that it matches anyone other than the defendant is 1 in 1,000,000,000. As one author wrote, "When you hear one in a billion . . . it's almost like the dead person sitting up, pointing and saying, 'He did it.'"[3]

Or perhaps the reasons for the phenomenon are based upon the national media providing potential jurors stories about the power of forensic DNA evidence. The national media have flooded the public with stories of death-row or long-term inmates being freed, almost instantaneously, through the use of DNA testing.[4] For instance, CNN reported in June 2000 that the State of New York had at least seven cases of convicted criminals being exonerated through the use of DNA testing.[5] Even more compelling is what happened in Illinois, where DNA testing led to the widely publicized release of Rolando Cruz, who served three years on Illinois' death row.[6] The use of DNA testing in the Rolando Cruz case and in the cases of other death-row inmates helped lead to a moratorium, comprehensive review, and ultimately a commutation of all death sentences in Illinois.[7]

Whatever the reasons for DNA's power, the forensic use of DNA in criminal proceedings enjoys an unmatched ability to convince jurors of guilt or innocence. It also has the same ability to convince judges, as was shown by the Indiana Supreme Court, in *Harrison v. Indiana*, where the court held: "the nature of the [DNA] expert testimony involves precise physical measurements and chemical testing, the results of which were not subject to dispute."[8] Because of these strong beliefs, forensic DNA presents a scientific tool that, as Jung said, can "work harm" for those whom it is used against. Consider, for example, the case of Lazaro Sotolusson, a man wrongly accused of two rapes. Sotolusson was charged after a lab technician mistakenly placed his name on another man's DNA profile.[9] DNA, as this example illustrates, works harm because it is susceptible to the same bias and error as the forms of forensic evidence that preceded it.

It is this harm that led a high-priced defense team to challenge the Los Angeles County's use of DNA evidence in the case against O. J. Simpson. No matter your view of the outcome of the O. J. trial, one must admit that the defense convincingly showed the inherent bias and errors in DNA testing.[10] Many criminal defendants, however, do not have the same ability to access such a defense and face the possibility of an ever wider use[11] of forensic DNA against them. Because of this wider use and the risk of harm, every defendant, regardless of means, needs access to a DNA expert when forensic DNA evidence will be used against him or her. Unfortunately, many states either have not allowed or have set high hurdles for criminal defendants to gain access to a government-funded DNA expert.[12]

THE RIGHT TO ASSISTANCE

It is a well-settled principle of law that a defendant, under the right set of circumstances, should be entitled to an "independent" scientific expert when expert testimony will be used to prosecute him or her.[13] This right springs out of the Fifth Amendment to the Constitution, which provides that no person "be deprived of *life, liberty*, or property, without the due process of law"[emphasis added]. Because of the constitutional right to due process, the courts have found that certain constraints can be placed on governmental decisions that "deprive individuals of 'liberty' or 'property' interests within the meaning of the Due Process Clause."[14] However, just showing that life or liberty is in jeopardy is not enough to gain access to one's own expert. The circumstances that lead to that decision involve the analysis of (1) whether a private interest is involved; (2) the risk of an erroneous deprivation of such interest, along with the probable value of additional safeguards; and (3) the government's interest.[15]

THE INTEREST

In *Ake v. Oklahoma* (1985), the Supreme Court found, as is most likely self-evident, that the loss of liberty is "uniquely compelling" in determining whether a private interest is at stake in a government proceeding. According to the Court, "The interest of the individual in the outcome of the State's efforts to overcome the presumption of innocence is obvious and weighs heavily in our analysis."[16] Thus, a criminal defendant who will have DNA forensic evidence used by the prosecution, and who faces a jail sentence or death penalty resulting from such prosecution, has a private interest.

RISK OF ERRONEOUS PROCEEDING
AND POSSIBLE SAFEGUARDS

There are a number of examples where the courts have found that the risk of an error exists when scientific testimony is used in the trial phase. The seminal case was *Ake v. Oklahoma*, where the scientific testimony being used was that of psychiatry. In *Ake*, the Supreme Court found that "without the assistance of a psychiatrist to conduct a professional examination on issues relevant to the defense to help determine whether the insanity defense is viable, to present testimony, and to assist in preparing the cross-examination of a State's witnesses, the risk of an inaccurate resolution of sanity issues is extremely high."[17]

Similar to *Ake*, without the assistance of a DNA expert to present testimony and assist in cross-examination of a state's witness, the risk of an inaccurate resolution of guilt or innocence is high. Defense counsel requires assistance in gaining a rudimentary knowledge of DNA and its role as a forensic tool. A lawyer is educated in the ways of the law and not the complicated nature of DNA. Without the rudimentary knowledge of the evidence presented, a lawyer will not understand the evidence nor how to challenge it. DNA evidence is presented to a jury using statistical probabilities—for example, the chance that the evidence from a crime scene matches anyone other than the defendant is x in y. The way that these statistical probabilities are presented tends to have vastly different effects on a jury. A study conducted at the University of Texas showed that most people "have poor intuitions when it comes to the reasoning with statistics in general."[18] Also, the study revealed that the way the statistics are presented will have vastly different outcomes on a juror's findings. Two examples from the study are as follows:

1. "The probability that the suspect would match the blood specimen if he were not the source is 0.1%."
2. "One in one thousand people in Houston who are not the source would also match the blood drops."

The study found that the majority of jurors were 99 percent certain the suspect was the source of the evidence in the first example and 99 percent certain he wasn't in the second example. This is amazing, considering that the statistical probabilities being presented are exactly the same. As this illustrates, DNA evidence if presented to a jury by an unchecked witness for the prosecution could lead to an inaccurate determination of guilt or innocence.

Further, the defense needs to be able to understand whether errors were conducted in the actual tests performed by the government. As is shown by the Sotolusson case, errors occur during testing. The most sound of scientific methods are only as good as the humans who apply them. The National Research Council summed it up best when it said: "There is no substantial dispute about the underlying scientific principles [of DNA]. However, the adequacy of laboratory procedures and the competence of the experts who testify should remain open to inquiry."[19] The exact error rate of these tests is not clear. However, one laboratory showed the error rates, using the old technique of Restriction Fragment Length Polymorphism (RFLP) testing, to be one in fifty.[20] Some studies surmised that these errors resulted from poor lab procedures, including the lack of blind testing. Many of the testing samples were sent to labs accompanied by the names of the suspect and the graphic details of the crime committed, which could have the obvious effect of swaying the examiner to find a match where one did not exist or to simply exaggerate the probability of a match.[21] The newer technique of Polymerase Chain Reaction (PCR) testing has eliminated some of these concerns; however, PCR testing is still subject to the easiest error of them all, the mixing of samples.[22] Whether the error rates are one in fifty or one in one hundred, as a study of British labs showed last year,[23] errors do occur. And, while they may occur only a small number of times, one error can be the difference between life and death for a criminal defendant.

A defense expert knowledgeable in the science of forensic DNA can assist a defense attorney in determining the best method for portraying the statistical probability of a match. The expert can also assist the defense attorney in conducting their own tests, which assure that the results of the government's tests are accurate. This expert will serve as a check to the prosecution's witness, much the same way a defense attorney serves as a check to the prosecution. This check will help guard against the improper loss of liberty or life.

THE GOVERNMENT INTEREST

Providing the defense DNA expert assistance primarily impacts the government's interests financially. The court in *Ake* found the financial impact of a single expert on the state to be minimal.[24] A survey of DNA experts has shown that the average review of forensic DNA evidence would take between four and six hours at an hourly rate of approximately $150, which is comparable to a sanity review by a psychiatrist.[25] And much the same as the psychiatrist, if the review shows that further tests are necessary, then the hours and costs would go up accordingly. Based on this, it is highly unlikely that a court is likely to find the financial impact of a DNA expert greater than that found in *Ake*.

Additionally, the government has a higher financial interest in seeing justice served properly during the trial phase. Consider the costs involved with the many previously convicted defendants who have been incarcerated at the state's expense for many years and who now are being set free.

A defendant who does not have the means should have access to a state-funded forensic DNA expert, but many states' current requirements do not support this. In Indiana, the state only requires that the forensic DNA expert be a "neutral" party.[26] Several post-*Ake* decisions at the Federal Appellate level have held that the defendant's expert must be the defendant's and independent of the state[27] and that "the essential benefit of having an expert in the first place is denied the defendant when the services of the doctor must be shared with the prosecution."[28] Or consider New Hampshire, where the requirement is a showing of necessity.[29] The *Ake* decision shows that *any* expert testimony used by the government could make a trial fundamentally unfair when the defense is left without its own expert.[30] Therefore, the very presence of DNA at the trial should qualify as a necessity. Without uniform equal access to a DNA expert for those who cannot afford it, this scientific tool will wind up causing the harm that Jung feared.

NOTES

1. Carl Jung, "Introduction" (1) to Commentary in Cary F. Baynes, trans., *The Secret of the Golden Flower* (New York: Causeway Books, 1961).

2. Jonathan J. Koehler, "The Psychology of Numbers in the Courtroom: How to Make DNA-Match Statistics Seem Impressive or Insufficient," *Southern California Law Review* 74 (2001): 1275, 1277.

3. See Harriet Chian, "Court Evidence Controversy: Jury Still Out on Effectiveness of DNA Analysis," *San Francisco Chronicle*, July 5, 1994, A1, noted in Koehler, "The Psychology of Numbers in the Courtroom," 1275, 1277.

4. See, for example, C. Thomas Blair, "Comment: *Spencer v. Commonwealth* and Recent Development in the Admissibility of DNA Fingerprint Evidence," *Virginia Law Review* 76 (1990): 853; Maurice Possley, "Prisoner to Go Free as DNA Clears Him in Beauty Shop Rape," *Chicago Tribune*, February 25, 1999.

5. "Capitol Hill Debates DNA Testing in Capital Crimes," *CNN.com*, June 13, 2000, available at robots.cnn.com/2000/ALLPOLITICS/stories/067/13/dna .hearing/.

6. "For Rolando Cruz and Alejandro Hernandez, the Third Time Was a Charm," available at www.IllinoisDeathPenalty.com/cruz.html.

7. "Illinois Governor Pardons Rolando Cruz, Two Others Wrongly Convicted of Murder," *SFGate.com*, December 19, 2002, www.sfgate.com/cgi.bin/article.cgi?file- /news/archive/2002/12/19/national1348EST0.

8. *Harrison v. Indiana*, 644 N.E.2d 1243, 1253 (Ind. 1995).

9. Glenn Puit, "Changes Proposed in DNA Handling," *Las Vegas Review-Journal* online edition, May 15, 2002, available at www.lvrj_home/2002/May-15-Wed-2002 /news/18751101.html.

10. See generally Ronald J. Allen, "The Simpson Affair: Reform of the Criminal Justice Process and Magic Bullets," *University of Colorado Law Review* 67 (1996): 989.

11. Laurel Beeler and William R. Wiebe, "Comment: DNA Identification Tests and the Courts," *Washington Law Review* 63 (1988): 903, 908 n. 21.

12. See generally *Cade v. Florida*, 658 So.2d 550, 555 (Fla. Dist. Ct. App. 1995); *Roseboro v. State*, 258 Ga. 39, 41(3)(d) (365 S.E 2nd 115) (1988) concurred in *Mosier v. State*, 218, Ga.App. 586, 587 (1995); *People v. Sims*, 244 Ill. App. 3d 966 noted in *Cade v. Florida*, 658 So.2d at 555; *Coleman v. Mississippi*, 697 So.2d 777, 782 (Miss. 1997); *State v. Stowe*, 620 A.2d 1023, 1027 (1993) noted in *Cade v. Florida*, 658 So.2d at 555; *State v. Mills*, 420 S.E.2d 114, 117-19 (N.C. 1992); *Ohio v. Kent*, No. 72435 at 15 (Ohio App. 1998) concurring in *State v. Scott*, 41 Ohio App. 3d 313, 315 (1987).

13. See generally *Ake v. Oklahoma*, 470 U.S. 68, 71 (1985); *Cowley v. Stricklen*, 929 F.2d 640 (11th Cir. 1991); *Smith v. McCormick*, 914 F.2d 1153 (9th Cir. 1990).

14. *Mathews v. Eldrige*, 424 U.S. 319, 332 (1976).

15. *Goldberg v. Kelly*, 397 U.S. 254, 260-71 (1970) noted in *Mathews v. Eldridge*, 424 U.S. 319, 325 n.4. (1976).

16. *Ake v. Oklahoma*.

17. *Ake v. Oklahoma*.

18. See, for example, Robin Hogarth, *Judgement and Choice*, 2nd ed. (Hoboken, NJ: Wiley, 1987), 15; Daniel Kahneman and Amos Tversky, "On the Psychology of Prediction," *Psychology Review* 80 (1973): 237; Michael J. Saks and Robert F. Kidd, "Human Information Processing Adjudication: Trial by Heuristics," *Law and Society Review* 15 (1980–1981): 123, 127. Referenced in Koehler, "The Psychology of Numbers in the Courtroom," 1275, 1277.

19. National Research Council, National Academy of Sciences, *DNA Technology in Forensic Science* (Washington, DC: National Academy of Sciences Press, 1992), 158. Quoted in Edward Connors et al., "Convicted by Juries, Exonerated by Science: Case Studies in the Use of DNA Evidence to Establish Innocence After Trial," *U.S. Department of Justice Research Report* 25 (June 1996).

20. Jay A. Zollinger, "Defense Access to State-Funded DNA Experts: Consideration of Due Process," *California Law Review* 85 (1997): 1817.

21. Craig M. Pease and James J. Bull, *DNA*, www.utexas.edu (2000), available at www.utexas.edu/course/bio301d/Topics/DNA/text.html.

22. Pease and Bull, *DNA*.

23. Nick Paton Walsh, "False Result Fear over DNA Tests," *Observer*, February 2002.

24. *Ake v. Oklahoma*, 78–9.

25. Zollinger, "Defense Access to State-Funded DNA Experts," 1833.

26. *Indiana v. Harrison*, 644 N.E.2d 1253.

27. *Cowley v. Stricklin*, 929 F.2d 640 (11th Cir. 1991).

28. *United States v. Sloan*, 776 F.2d 926, 929 (10th Cir. 1985).

29. *State v. Stowe*, 620 A.2d 1023, 1027 (1993) noted in *Cade v. Florida*, 658 So.2d at 555.

30. See *Ake v. Oklahoma*.

Part X

PRENATAL GENETIC MODIFICATION

"All people have the right to have been conceived, gestated, and born without genetic manipulation."

Article 10 of the Genetic Bill of Rights

· *24* ·

The Perils of Human Developmental Modification

Stuart A. Newman

*A*rticle 10 is the most unusual of all the precepts in the Genetic Bill of Rights in that it appears to incorporate a paradox. What does it mean for someone to have the right not to have been brought into existence under conditions that, had they been different, might have resulted in someone else? It also can be read (incorrectly, I will argue) as attributing rights to embryos and fetuses, clearly not the intention of the drafters of this document.

The indispensability of Article 10 to the Genetic Bill of Rights can be seen when it is considered in relation to the unexceptionable and uncontroversial Article 5: "All people have the right to protection from toxins, other contaminants, or actions that can harm their genetic makeup and that of their offspring." But if adults have the right not to be exposed to toxins, and they have the right for their offspring not to be exposed to toxins, what about the offspring themselves? It is not much of a stretch to assert from this that they (that is "all people," since we were all once there) have the right to have been "conceived, gestated, and born" without being exposed to toxins. This is not an attribution of rights to embryos and fetuses; it clearly pertains to people who are full persons.

If a business or municipality dumps toxins into drinking water, the wrong it does extends not only to existing people in their own right, but also to people in their capacity as future parents of unhealthy children. Moreover, that right to protection against such harms does not just apply to existing persons; the future children themselves, who may suffer impairments to their own bodies and health, are also wronged. Similarly, if a pharmaceutical company sells an inadequately tested product and this causes disability or illness to the offspring of a pregnant woman who consumes it, the wrong, both in the legal sense and in common understanding, is not just to the woman in her capacity as a parent but also to the disabled or ill child born to that woman. The case of diethylstilbestrol (DES), prescribed to pregnant women in the 1950s and 1960s, under the assumption (incorrect, as it turns out) that it prevented miscarriages, and which caused cancers of the reproductive system in many of their daughters exposed to it in utero is one example. Another is thalidomide, a sleeping pill prescribed for pregnant mothers in the 1960s, which caused severe limb reductions and other developmental anomalies in their children.

With regard to chemical toxins, there is a gray area with respect to rights and responsibilities that it would be disingenuous not to consider in some detail. This relates to the role of the mother as a conduit, or even administrator, of the toxic substance to her developing child. At one end of this range of situations is the case where a mother requires an embryotoxic medication for her own survival. Certain anti-seizure medications taken by epileptic patients, for example, are well known to adversely affect embryonic development. At the other end are voluntary activities such as smoking, which leads to low-birth-weight children with numerous associated health problems, and alcohol and recreational drugs, many of which are embryotoxic. In between are other risky behaviors, some of which may be all but involuntary, such as drug use by the addicted, or being poor and living near a toxic dump site.

Some of these considerations have been marshaled by opponents of women's reproductive and personal autonomy to stigmatize and criminalize behaviors that may jeopardize an unborn child. The most extreme of these is, of course, abortion, the elimination of the right to which is the ultimate objective of these groups. It seems likely that the only thing that has restrained an even more aggressive campaign by the political Right against the autonomy of pregnant women is the associated presumption of liability on the part of corporate polluters and poisoners.

Of course, one can oppose criminalization of a person's activities with respect to her own body while at the same time being critical of voluntary activities that may harm a developing embryo or fetus intended to be brought to term. In particular, if one advocates the legal requirement for the liquor and pharmaceutical industries to warn about the adverse effects of their products, it is untenable to be completely neutral concerning a pregnant woman's failure to heed the warnings.

With regard to the genetic engineering of future children, there is no such gray area. Embryo gene manipulation is untested in humans. Unlike cases in which a woman needs to be medicated despite potential dangers of her medicine to her unborn child, there is no compelling need to administer new genes to an embryo intended to be brought to term. As stated in the Council for Responsible Genetics position paper on human germline gene manipulation, such procedures are "not needed in order to save the lives, or alleviate the suffering, of existing people. Its target population are 'prospective people' who have not even been conceived."

The hazards of genetic modifications to humans have usually been discussed in terms of *somatic* (body cell) modification, in which only nonreproductive tissues are affected, and *germline* (egg or sperm cell) modification, in which changes to an individual's DNA can be passed down to future generations. But genetic modification of early embryos (similarly to cloning) subjects the developing individual to hazards even if there is no germline transmission to future generations.

The dangers of germline transmission of DNA modification are clear. For example, germline introduction of an improperly regulated normal gene resulted in progeny of the modified animal with no obvious effects on development, but enhanced tumor incidence during adult life.[1] Such effects may not be recognized for a generation or more.

It is important to recognize, however, that hazards of such alterations are not eliminated when there is no germline transmission. The biology of the developing individual may still be profoundly altered by the manipulation of his or her genes at an early stage. Laboratory experience shows that insertion of foreign DNA into unfavorable sites in an embryo's chromosomes can lead to extensive perturbation of development. For example, the disruption of a normal gene by insertion of foreign DNA in a mouse caused abnormal circling behavior when present in one copy, and lack of eye development, lack of development of the semicircular canals of the inner ear, and anomalies of the olfactory epithelium, the tissue that mediates the sense of smell, when the mice were inbred so that the mutation appeared in the homozygous form (i.e., on both copies of the relevant chromosome).[2]

Another such "insertional mutagenesis" event led to a strain of mice that exhibited limb, brain, and craniofacial malformations, as well as displacement of the heart to the right side of the chest, in the homozygous state.[3] Each of these developmental anomaly syndromes was previously unknown. From current, or even anticipated, models for the relationship between genes and organismal forms and functions, the prediction of complex phenotypes on the basis of knowledge of the gene sequence inserted or disrupted is likely to remain elusive. It is quite clear, then, that the first time or the first hundred times a gene insertion or replacement is attempted with the intention of bringing the manipulated embryo to term ("developmental" gene modification), it will constitute unwarranted experimentation on human beings, in violation of the fifth provision of the 1949 Nuremberg Code governing human experimentation, which states: "No experiment should be conducted, where there is an *a priori* reason to believe that death or disabling injury will occur; except, perhaps, in those experiments where the experimental physicians also serve as subjects."[4]

While unsuccessful attempts at developmental gene modification can profoundly perturb ordinary biological function and introduce new, harmful genetic variants into the gene pool, even "successful" attempts would be problematic. Few scientists or knowledgeable observers believe that the objective of prospectively curing genetic conditions would remain the primary medical agenda. This is because, with rare exceptions, prenatal and pre-implantation diagnoses are sufficient for identifying and selecting embryos free from genetic variants the parents do not wish to transmit. So the focus would then shift to introducing gene variants into developing embryos different from either parent, with the aim of producing biologically customized and eventually "better" people.

Although futurologists and some bioethicists have applauded the prospect of such ventures into what critics have termed "Yuppie Eugenics,"[5] a societal division into genetic haves and have-nots is the likely outcome, to the extent that experimental failures among the offspring of technophilic "early adopters" can be controlled and the results can be made to generally reflect the parents' expectations. But the genetic have-nots will not only be those denied the "good" genes; among them will be those prospective star athletes and geniuses inadvertently damaged by their parents' desire for the bigger and better. Whereas eighteenth-century kapellmeisters could order up biologically manipulated *castrati* to satisfy the requirements of their choral ensembles, some achieving legendary status, the lives of the failed experiments, unable or unwilling to play their foreordained roles but mutilated nonetheless, are lost to history.

Prospective developmental manipulation, twenty-first-century style, for the purpose of giving a child talents and other favorable traits, would not be so different from the eighteenth-century version. Unlike biological manipulations undertaken after development is complete—cosmetic surgery, heart valve and joint replacement, and even "somatic" (differentiated body cell) gene therapy—prepuberty castration and embryo gene manipulation both change the generative trajectory of the individual, turning it into something intrinsically different from what it would have become without the manipulation. With gene- and chromosome-based procedures, there is no guarantee that even the original species character will be maintained. Although such methods may be undertaken to fabricate improved humans, in some cases, by accident or by intent, the outcomes will be quasi-human or less than human.

Or the manipulation may just lead to painful consequences. A study in which a mouse embryo was genetically modified with the intention of inducing a changed behavioral profile accomplished this goal: the mature mouse performed in a superior fashion on several tests of learning and memory[6] and was featured in the popular media for a while as the "Doogie" mouse, after a fictional child prodigy. What was not as widely reported was that these mice also exhibited enhanced sensation of pain when exposed to chronic stimuli.[7]

Contrary to popular misconceptions (often abetted by scientists of a reductionist bent and uncritical journalists), genes do not constitute an organism's "blueprint" or "program"; the genotype determines the phenotype in only an approximate sense.[8] A study that compared outcomes of behavioral tests on inbred, genetically uniform strains of mice conducted in three different laboratories showed systematic differences across environments that were designed to be identical.[9] The researchers concluded that effects of a given genetic alteration on behavior could differ markedly despite uniformity of genetic background and setting.

Humans, of course, are much less genetically uniform than inbred strains of mice, and it is to be expected that many, if not most, attempts at genetically engineering children will have unexpected adverse outcomes. One way of bringing such uncertainty under partial control (to follow the logic of this questionable enterprise) is to start with hundreds of cultured embryo stem (ES) cells produced by cloning from a known prototype, and attempt to correct or improve on the prototype. But then, the ideology of the new and improved, which motivates this whole line of activity, would work against the acceptance of the inevitable experimental errors—children with brain damage and other profound disabilities resulting from genetic engineering gone awry—spurring parents in search of perfection to try again for a better result. In effect, the quality control paradigm appropriate to any design-oriented technology would set in, and for the first time production of human beings will enter the realm of manufactured items.

As with any harm due to early, damaging interventions—teratogenic insult by toxic or viral agents, random genetic mutation, childhood physical or psychological abuse—it is no disrespect to the affected individual to wish that it had not happened and to try to prevent it from happening to others. This is even the case for interventions made in good faith, such as DES treatment for pregnant women. At a moment in history when developmental gene manipulation has not yet been attempted in humans, but when the technical capacity is in place and it is increasingly being called for by hucksters and philosophers alike, it is no affront to anyone who (notwithstanding the counsel of the "Precautionary Principle"[10]) may be produced this way in the future, nor is it an incursion on any woman's reproductive autonomy (which never included the right to produce children by any means possible), to affirm that all people have the right to have been conceived, gestated, and born without genetic manipulation.

NOTES

1. A. Leder, P. K. Pattengale, A. Kuo, T. A. Stewart, and P. Leder, "Consequences of Widespread Deregulation of the c-myc Gene in Transgenic Mice: Multiple Neoplasms and Normal Development," *Cell* 45 (1986): 485–95.

2. A. J. Griffith, W. Ji, M. E. Prince, R. A. Altschuler, and M. H. Meisler, "Optic, Olfactory, and Vestibular Dysmorphogenesis in the Homozygous Mouse Insertional Mutant Tg9257," *Journal of Craniofacial Genetics Developmental Biology* 19 (1999): 157–63.

3. G. Singh et al., "Legless Insertional Mutation: Morphological, Molecular, and Genetic Characterization," *Genes and Development* 5 (1991): 2245–55.

4. In *Trials of War Criminals Before the Nuremberg Military Tribunals Under Control Council Law No. 10* (Washington, DC: U.S. Government Printing Office, 1949), 181–2.

5. Ruth Hubbard and Stuart A. Newman, "Yuppie Eugenics," *Z Magazine* (2002): 36–9.

6. Y. P. Tang et al., "Genetic Enhancement of Learning and Memory in Mice," *Nature* 401 (1999): 63–9.

7. F. Wei et al., "Genetic Enhancement of Inflammatory Pain by Forebrain NR2B Overexpression," *Nature Neuroscience* 4 (2001): 164–9.

8. Stuart A. Newman, "Idealist Biology," *Perspectives in Biology and Medicine* 31 (1988): 353–68; Richard C. Strohman, "Ancient Genomes, Wise Bodies, Unhealthy People: Limits of a Genetic Paradigm in Biology and Medicine," *Perspectives in Biology and Medicine* 37 (1993): 112–45; L. Moss, *What Genes Can't Do* (Cambridge, MA: MIT Press, 2003).

9. J. C. Crabbe, D. Wahlsten, and B. C. Dudek, "Genetics of Mouse Behavior: Interactions with Laboratory Environment," *Science* 284 (1999): 1670–2.

10. C. Raffensperger and J. Tickner, eds., *Protecting Public Health and the Environment: Implementing the Precautionary Principle* (Washington, DC: Island Press, 1999).

Human Rights in a Post-human Future

Marcy Darnovsky

$\mathcal{M}$ost people are well aware that efforts to "improve the human gene pool" and "breed better people," notoriously widespread from the end of the nine-teenth century through the middle of the twentieth century, led to some of the most extreme violations of civil, political, and human rights in recent history. Nonetheless, five or six decades ago—before the structure of DNA had been deduced, before the modern environmental movement—most of the provi-sions of the Genetic Bill of Rights would have seemed nonsensical.

Even twenty-five years ago—before the development of genetic manipu-lation at the molecular level, legal doctrines that allow governments to grant patents on life, and DNA databases; before the advent and commercialization of in vitro fertilization and the screening of in vitro embryos; before the appear-ance of advertisements for social sex selection in mainstream U.S. publications—the document would have been widely considered an unwarranted over-reaction based on dystopian fantasy.

But here we are, at the beginning of the twenty-first century. Plants and animals are routinely genetically modified, patented, and brought to market by corporate enterprises. Genetic technologies are increasingly applied to human beings for forensic and medical purposes. The biotechnology industry, though it has lost over \$40 billion since its inception twenty-five years ago,[1] continues to attract large amounts of venture capital and generate glowing headlines.

As many have observed, public understanding of these trends is lagging far behind technical developments and commercial deployment. Many people feel daunted by their technical complexity, and therefore reluctant to make politi-cal or ethical judgments about them. Grappling with the social meaning of the various human genetic technologies has proven thorny even for those whose political commitments usually make them wary of corporate-dominated technological projects.

Though the environmental movement has gained at least a toehold for a precautionary approach to powerful new technologies, this principle is often disregarded when technical innovations are presented as medical advances. And that is a move that the biotechnology industry and its supporters have mastered. Their claims about the future of human genetic technologies are nothing if not ambitious. Revolutions in health care are invoked. Promises of imminent medical miracles proliferate. Technical fixes for global health inequities are proposed and funded. Senior researchers are unembarrassed to suggest that aging, even death, can be overcome through biotechnological engineering.[2]

REPRODUCTIVE GENETICS AND THE POST-HUMAN AGENDA

The most troubling of the human biotechnologies are those that involve reproduction. Currently, the procedure known as pre-implantation genetic diagnosis (PGD) allows the screening and selecting of embryos on the basis of sex and other traits. Many feminists and disability rights advocates are deeply uneasy about this practice. But for some enthusiasts, such crude forms of selection are just the beginning. More than a few observers predict that in vitro embryos will one day be manipulated and modified rather than merely screened and selected. They point out that the genetic technologies now being used routinely to alter mammalian species, if applied to humans, would permit the "redesign" of the traits of future children.

Some proponents of inheritable genetic modification (IGM) predict that within a generation "enhanced" babies will be born with increased resistance to diseases, optimized height and weight, and increased intelligence. Farther off, but within the lifetimes of today's children, they foresee the ability to adjust personality, design new bodily forms, extend life expectancy, and endow hyper-intelligence.[3]

Eagerness to provide parents with the technical means to redesign their future offspring is often coupled with a larger social vision. Advocates of IGM point out that manipulating the genetic makeup of future generations amounts to "seizing control of human evolution."[4] They correctly observe that coupling the techniques of inheritable genetic modification with existing social and market dynamics could trigger a self-reinforcing spiral of eugenic engineering, perhaps culminating in the abandonment of our common biological identity as human beings. Some anticipate a "post-human" future, called into existence through consumer choices in a market-based eugenics, and the subsequent emergence of "genetic castes."[5]

Is such a future likely? Hopefully, scenarios like these will remain beyond technical reach. Notwithstanding the flesh-and-blood accomplishments of to-day's genetic scientists—glow-in-the-dark rabbits, goats that lactate spider silk, and the like—modified genes and artificial chromosomes may never work re-liably. Transgenic designer babies may be too ridden with unpredictability or malfunction to ever become a popular option.

But both the trajectory of human biotechnology and the growing ideo-logical influence of high-biotech libertarian futurism counsel that we take these visions seriously. After all, their purveyors are not limited to the marginal "cowboy cloners" and others on the far shore of credibility.[6] Also among those who eagerly anticipate a post-human future are congeries of biomedical re-searchers, biotech entrepreneurs, bioethicists, and other scholars. A disturbing number of them are respected figures working at prestigious institutions and wielding significant cultural influence. Also disquieting is the near silence of their scientific colleagues. Many of them must have qualms about the use of biotechnology in the service of a new eugenics, but few have publicly reg-istered concern.[7]

A HUMAN RIGHTS FRAMEWORK
FOR REPRODUCTIVE GENETICS

It is at this historical moment that a small U.S.-based non-governmental or-ganization has written and proffered the Genetic Bill of Rights. Like other declarations of rights, this one makes bold claims about the social conditions that characterize our world, and about those that should. The Genetic Bill of Rights asserts the profound consequentiality of new knowledge in the genetic sciences and new techniques of biological manipulation, and of the legal and commercial contexts in which they are being developed and deployed. And it asserts the urgency of establishing a broad consensus about how the new knowledge and technologies should be governed.

In a landmark article titled "Protecting the Endangered Human Species: Toward an International Treaty Prohibiting Cloning and Inheritable Alter-ations," published in the *American Journal of Law and Medicine*, George Annas, Lori Andrews, and Rosario Isasi argue that the human condition of belonging to a single biological species is "central to the meaning and enforcement of hu-man rights." Because reproductive cloning and inheritable genetic modifica-tion "can alter the essence of humanity itself," they write, these techniques "threaten to change the foundation of human rights." For this reason, the au-thors say, "cloning and inheritable genetic modification can be seen as crimes against humanity of a unique sort."[8]

In many parts of the world outside the United States, the technologies of human genetic redesign are commonly and comfortably viewed through the lens of human rights. These procedures would be easily and widely understood as violations of what the Genetic Bill of Rights calls "the right to have been conceived, gestated, and born without genetic manipulation." Especially in light of the fact that neither reproductive cloning nor inheritable genetic modification has yet been applied to human beings, the strength of the sentiment for national and international bans on them is striking.

Some dozens of countries have already passed such bans, and several important multilateral instruments address these technologies under the rubric of human rights.[9] The Council of Europe, for example, prohibits both inheritable genetic modification and human reproductive cloning in its Convention on Human Rights and Biomedicine, which was opened for signatures in 1997 after several years of negotiations and preparations.[10] Similarly, UNESCO's Universal Declaration on the Human Genome and Human Rights, though not a legally binding document, forbids the production of cloned human beings, and says that inheritable genetic modification "could be contrary to human dignity."[11]

LIBERTY AND JUSTICE IN THE AGE OF GENETICS

In the United States, the claim of a right "to have been conceived, gestated, and born without genetic manipulation" resonates less strongly. We tend to think of "rights," including "human rights," as shielding individuals from the coercive power of the state. However, today commercial entities often have as much control over individuals' life choices and destinies as do governments. If the biotechnology and assisted reproduction industries were to decide to develop "genetic enhancement" procedures and market them to prospective parents, the pressures to "provide the best start in life for your child" would be considerable. Health insurance companies would likely weigh in. Coercion of parents need not necessarily be enforced by governmental authority to be effective. And the children in question, of course, would have no protection—unless we establish that freedom from genetic manipulation is indeed a human right.

In the individualist culture of the United States, rights are usually meant first and foremost to protect and enlarge individual liberties. The discourse of human rights, by contrast, implies as well the imperative of safeguarding the collective conditions in which people and communities can flourish. In the United States, we tend also to characterize rights as applying to us as autonomous beings who choose our own values and chart our own lives, rather than as people unavoidably situated in complex and overlapping relationship with each other. As autonomous individuals, we go it alone. As social beings necessarily dependent on each other in myriad fashion, we are obligated to struggle together toward shared understandings about the kind of world we want to build.

Finally, we in the United States often focus so sharply on individual rights and liberties that we blur our perception of the social conditions that foster or block their enjoyment. We sometimes forget, in other words, that rights are necessarily embedded in relations of power. But championing rights in the abstract, without considering the political and social inequities with which we live, can undermine our commitments to social justice and solidarity, and to the democratic principle that we can and ought to participate in decisions about the basic conditions of our polity and collective life.

In her investigation of the tension between individual liberty and social justice as it pertains to reproductive rights and racial equality, legal scholar Dorothy Roberts asserts that the "dominant view of liberty reserves most of its protection only for the most privileged members of society." By contrast, she argues that "reproductive freedom is a matter of social justice," and that "pro-creation's special status stems as much from its role in social structure and political relations as from its meaning to individuals." She is appalled that advocates of the new eugenics can present themselves as champions of freedom even as they "dismiss the possibility that genetic enhancement might exacerbate race and class disparities."[12]

Liberties and rights, no matter how loudly they are proclaimed to be "self-evident," are always the results of social negotiations, often painfully arrived at, on matters of common concern. Most nations of the world have now abolished slavery. Many have criminalized marital rape and outlawed the selling and abuse of children. These are examples of widely accepted limits on practices once construed as rights.

In practice, the two conceptualizations of rights—call them the individual-choice-and-autonomy model and the social-justice-and-negotiation model—often co-exist in the same policy formulation. For example, the right not to be enslaved protects individuals from being subjected to involuntary servitude, yet the same right also bespeaks a socially negotiated—albeit once hotly contested—agreement that a world in which some have the power to enslave others is not a world in which we wish to live.

GENETIC RIGHTS AND WRONGS

How does all this apply to the proposed "right to have been conceived, gestated, and born without genetic manipulation"? Advocates of market-based eugenics, appealing to the widely accepted consumer-oriented norms of our society, and to the very high value it places on individual liberty, scientific free-dom, and technological advance, argue that people have the right to select the traits of their future children. Often they present this as an extension of repro-ductive choice and "procreative liberty."[13]

These assertions can be countered even from within the individual-choice-and-autonomy model of rights. Experience with cloned and transgenic animals demonstrates that such procedures would carry enormous risks for both the cloned or genetically modified child and for the child's mother. As developmental biologist Stuart Newman points out, "no amount of data from laboratory animals will make the first human trials anything but experimen-tal." And since there is little medical justification for such procedures, they would represent a clear-cut case of unethical human experimentation.[14]

Furthermore, it would be impossible to obtain what bioethicists call "in-formed consent" from the person to be cloned or modified, since the proce-dure would have to be carried out well before birth. And reproductive cloning and inheritable genetic modification would arguably compromise the auton-omy of the cloned or modified person, since his or her life would have been controlled in an unprecedented manner by the parents, fertility doctors, and biotech companies involved.

The social-justice-and-negotiation model of rights provides additional support for the proposed right to be born free of genetic manipulation. It at-tends with care to the likelihood that the commercial development of repro-ductive cloning and IGM would exacerbate existing inequalities and create new forms of discrimination and inequality. It heeds the dangers of granting novel forms of control over individuals' lives, and over the genetic legacy of the human species, to any public or private entity.

The Genetic Bill of Rights, and the "right to have been conceived, ges-tated, and born without genetic manipulation" that it contains, is a statement of political will and moral intelligence. In an era that has witnessed dire con-sequences of technological grandiosity, it calls for extending the precautionary principle to our own biology. In an age of runaway elitism, it calls for affirm-ing our common humanity as a minimal but crucial condition of solidarity and mutuality. In the face of efforts to inscribe inequality into the human genome, it insists that—like it or not—we're all in this together.

NOTES

1. David P. Hamilton, "Biotech's Dismal Bottom Line: More Than $40 Billion in Losses," *Wall Street Journal*, May 20, 2004.

2. Stephen S. Hall, *Merchants of Immortality: Chasing the Dream of Human Life Extension* (Boston: Houghton Mifflin, 2003).

3. See LeRoy Walters and Julie Gage Palmer, *The Ethics of Human Gene Therapy* (New York: Oxford University Press, 1997); Gregory Stock and John Campbell, *Engineering the Human Germline: An Exploration of the Science and Ethics of Altering the Genes We Pass to Our Children* (New York: Oxford University Press, 2000).

4. Stock and Campbell, *Engineering the Human Germline.*

5. Perhaps the best-known statement of this vision in a lay book is Princeton biologist Lee M. Silver's *Remaking Eden: Cloning and Beyond in a Brave New World* (New York: Avon Books, 1998).

6. For a short account of claims by Severino Antinori, Panos Zavos, and the Raelians that they are actively trying to clone a human being, see www .genetics-and-society.org/analysis/promodeveloping/cloning.html.

7. Among senior genetic researchers, a rare exception is Whitehead Center for Genome Research Director Eric Lander's short essay, "In Wake of Genetic Revolution, Questions About Its Meaning," *New York Times*, September 12, 2000. "The hardest question is, To what extent will we decide to reshape the genes we pass to our children? Some of my close colleagues are already proposing ways to 're-engineer' what they view as an 'imperfect' human genome . . . by modifying the human germline. . . . I part company from some of my colleagues here. While I'm strongly opposed to laws limiting scientific investigation, I would support a ban on modifying the human germline."

8. George Annas, Lori Andrews, and Rosario Isasi, "Protecting the Endangered Human: Toward an International Treaty Prohibiting Cloning and Inheritable Alterations," *American Journal of Law and Medicine* 2, nos. 2/3 (2002): 151–78.

9. Countries that have banned reproductive cloning or IGM include Australia, Austria, Argentina, Belgium, Brazil, the Czech Republic, Costa Rica, Denmark, France, Germany, India, Israel, Italy, Japan, Lithuania, Mexico, the Netherlands, Norway, Peru, Portugal, Romania, Russia, Slovakia, South Africa, South Korea, Spain, Sweden, Switzerland, Trinidad and Tobago, and the United Kingdom. For a complete list, see www.genetics-and-society.org/policies/other/index.html.

10. Relevant passages are excerpted at www.genetics-and-society.org/policies/ international/council.html. The Convention on Human Rights and Biomedicine, conventions.coe.int/treaty/EN/searchsig.asp?NT=164. The Additional Protocol [on human cloning], conventions.coe.int/treaty/EN/searchsig.asp?NT=168.

11. Relevant passages are excerpted at www.genetics-and-society.org/policies/ international/unesco.html. The Universal Declaration on the Human Genome and Human Rights, www.unesco.org/human_rights/hrbc.htm.

12. Dorothy Roberts, *Killing the Black Body: Race, Reproduction, and the Meaning of Liberty* (New York: Random House, 1997).

13. John Robertson, *Children of Choice: Freedom and the New Reproductive Technologies* (Princeton, NJ: Princeton University Press, 1994); Allen Buchanan et al., *From Chance to Choice: Genetics and Justice* (Cambridge: Cambridge University Press, 2002).

14. Stuart A. Newman, "The Hazards of Human Developmental Gene Modification," *GeneWatch,* July 2000, www.gene-watch.org/newman.html.

· *26* ·

Rights for Fetuses and Embryos?

Ruth Hubbard

$\mathcal{T}$he notion of a "right to be conceived, gestated, and born without genetic manipulation" runs counter to women's rights to reproductive autonomy—which the Council for Responsible Genetics (CRG) supports—by pitting embryonal and fetal rights against women's reproductive rights. How each of us is "conceived, gestated, and born" depends on the decisions made not by us, but by our future parents and especially our future mothers at a time when we ourselves either do not yet exist in any form or, at least, not in a form to which reproductive rights advocates grant either autonomy or rights that trump the rights of the woman who will be responsible for our conception, gestation, or birth.

If we delete the phrase "without genetic manipulation," the statement makes no sense at all. There is no right to be conceived, and depending on the legal status of abortion, a conceptus may or may not be considered to have a right to be gestated and born. In the United States so far, fortunately, a fetus does not have this right. So, does the phrase "without genetic manipulation" change this situation? Does it mean: "A person who is going to be conceived, gestated, and born has the proactive right to have gone through these processes 'naturally,'" whatever that means?

What does it mean in practice? Does introducing a sperm cell into an egg cell in vitro constitute genetic manipulation? If not, why not? And if so, then does in vitro fertilization offend against this right, especially if it involves intracytoplasmic sperm injection (ICSI)? Further, does this right empower the state to prevent a pregnant woman from engaging in any behavior that might affect the genetic constitution of her fetus? And, indeed, what behaviors might not do so—drinking coffee, taking certain medicinal or recreational drugs, working in hospitals or other potentially genotoxic environments? This right is incoherent and makes no sense.

If we were to take this right seriously, how would one litigate against its denial? The closest analogy I can think of would be wrongful death and wrongful life (or wrongful birth) suits. Wrongful death suits are suits in which the parents of a fetus or child who is believed to have died as a result of medical negligence or mistakes sue the professionals or hospital deemed to be responsible. Such suits are much like other malpractice suits and are adjudicated in the usual way. Wrongful life or wrongful birth suits are more unusual because there the child must establish legal standing to bring the suit. (Under U.S. law, a fetus does not have legal standing, nor does an embryo.) So far, no such suit has been won because judges generally hold that, no matter what one's state of health, it is preferable to have been born than not to have been born. Hence, no life or birth is held to be "wrongful."[1]

By analogy, it seems likely that, if someone were to litigate having had his or her genome "manipulated" before birth, the medicosocial climate would favor the experts who would argue that the procedure had been medically justified. It is unlikely to be a winnable situation and certainly not one that warrants calling women's reproductive rights into question. Furthermore, if fetal "gene therapy" ever came to be practiced, who would rightfully be able to protect a fetus against it? Surely not the fetus itself. So, who would be in a position to enforce this right?

A perhaps somewhat comparable situation involves infants who are born with ambiguous-looking genitalia that do not permit an immediate, clear-cut assignment of sex. Until recently in the United States, when a baby was born with ambiguous genitalia or with discordances between chromosomal and anatomical indicators such that pediatricians had trouble deciding whether the infant was male or female, the physician would decide which external markings of sex were easiest to ablate, modify, or construct surgically and reassure the parents that they were merely helping to reveal the child's true sex. The parents would then be instructed to rear the surgically modified child in a manner congruent with the medically assigned sex.[2]

In the past ten or so years, however, some of the people who had been manipulated in this way have organized the Intersex movement, analogous to the Gay and Lesbian or the Transsexual movements. And recently, they have begun to find allies among pediatricians and pediatric surgeons, who are ready to counsel and support the parents as they try to help intersex children accept their differences. These physicians agree that no hormonal or anatomical changes should be initiated until the children are old enough to decide whether, and in what way, their anatomical or physiological sex should be modified.[3]

Though this analogy may seem far-fetched, it can be used to address the question of who has the right to decide whether, and in what way, it is permissible to modify a person's genome. Such pseudo-analogies not withstanding, I continue to believe that the negative impact of this supposed right far outweighs its potential benefits. If we are to retain something like it, I would suggest rewording it as follows: "All people have the right to the integrity of their genome and to refuse to participate in attempts to modify it for any reason or at any point in their lives."

This phrasing does not explicitly cover the prenatal period, but it could, by implication, be stretched to do so by the parents to be. I do not think we are politically or legally justified to go any further.

NOTES

1. Ruth Hubbard, "The Politics of Fetal-Maternal Conflict," in Gita Sen and Rachel Snow, eds., *Power and Decision: The Social Control of Reproduction* (Cambridge, MA: Harvard University Press, 1994), 311–24.

2. Anne Fausto-Sterling, *Sexing the Body: Gender Politics and the Construction of Sexuality* (New York: Basic Books, 2000).

3. Alice Domurat Drager, ed., *Intersex in the Age of Ethics* (Hagerstown, MD: University Publishing Group, 2000).

Afterword: Focusing Ingenuity with Human Rights

The amendments to the Constitution of the United States that are known as the Bill of Rights have been studied and interpreted for more than two hundred years. They continue to evolve, to be individualized, particularized, recast, and enforced in ways reflecting changes to both the citizens and societies of this country. Their meaning, power, and the protections they confer are in part derived by their primacy in law, but also by the personal relationship all citizens have with them. They are, in fact, interpreted by judges but reinvented by each of us, throughout our lives, in the country that adopted them.

The issues and rights responses that are required for governance of biotechnology are similarly demanding and singularly personal. For me, developments in the life sciences pose important challenges to identity and personhood. They impact roles we play as men and women; as husbands, wives, and children; as family members; in reproduction and parenting; as members of groups and institutions; as sick and well individuals; as neighbors, community members, and citizens; and as workers and professionals. They may simultaneously alter us as individuals, change our societies, and require restatements of traditional rights, as well as consideration of new ones.

Consider the striking changes that biotechnology is bringing about:

- *Conception outside the body.* In-vitro fertilization and its first success, Louise Brown, are twenty-five years old, and on the horizon is gestation outside of a woman's body. Reproductive techniques have already created new families, mixing gamete donors, gestation surrogates, and parents in previously unimagined ways.

- *Selection of genetic effects on human traits prior to implantation of an embryo or during pregnancy.* The selection can be motivated not only by a desire for perceived beneficial effects, but also for utilitarian purposes, such as post-birth organ or tissue transplantation.
- *Enhanced testing of newborns.* In addition to screening for illnesses, this testing will be expanded for developmental, cognitive, behavioral, and achievement-associated traits. Soon this may result in lifelong medicalization and risk modification. A harbinger of this development is the chronic use of Ritalin and psychotropic medications for behavioral modification in children.
- *Enhancement of humans.* Now accomplished primarily by surgical methods and hormone treatments, genetic selection and manipulation will soon be considered for this end. In addition, implanted machines and nanotechnologies are being designed to supplement—or curtail—our thoughts and memories.
- *Life prolongation.* We now prolong life with therapies that include organ and tissue transplantation; over the last several decades, bodies have become chimeric—part animal or machine and part human. Soon scientists may attempt to directly manipulate genes that impact cellular aging.
- *Group identification.* Genetic tests are creating new associations, layered upon others that include race, ethnicity, gender, sexual orientation, religion, and location. In the future, tests could be used to choose our mates and neighbors.
- *Identity testing and profiling.* Governments and businesses currently use genetic technologies for their own purposes, such as DNA identification. Workplace genetic profiling for health and attitude issues may soon be attempted.

How should we respond? These new conditions challenge us to both reexamine and re-assert enlightenment values, to articulate and make effective rights, both old and new, like those discussed in this book. They provide an opportunity to conceive and enforce new policies that balance old truths with new knowledge and techniques and with the benefits of new technology for those in need. There are many traditions that need reconsideration and reformulation:

- *The right to a private space protected against involuntary biological and medical surveillance and analysis.*
- *The right to nondiscrimination and nonprofiling on biological grounds* (chapters 19, 20, and 21).
- *The right to technology that defends our individual freedoms and rights,* such as those that have aided in the release of wrongly convicted people

(chapters 22 and 23) and those that have been used to identify human remains in the aftermath of genocide.

- *The right to control personal medical information and gain access to it*—to selectively share it with others, such as health providers and family members, and to learn and benefit from these relationships (chapters 17 and 18).

Human creativity and ingenuity can produce new technology. Once created, these novel developments can be applied in differing contexts with varying impacts. Most will be ineffective, will create no appreciable change, and will not be worth any further public or private investment. Occasionally, important impacts result from new technology, but depending on the context, they may be more likely to cause harm—through poor individual or public outcomes or increased user dissatisfaction—than to produce salubrious effects. Rarely, technology innovations can be "disruptive" and participate in major changes to individuals and environments in which they act.[1]

No matter what the assessment of the value of new technology is at any particular time, its study may provide insight and perspective on important, interacting social, cultural, and political factors. It is clear that the push to create and apply biotechnology arises from complex personal and social phenomena. The difficulties that life poses for many people from birth and the wide extent of human suffering that might be ameliorated by human ingenuity should never be underestimated as a motivation. But in addition to attempts to address these human concerns, biotechnology has provided insight and illumination into challenging issues, old and new. Its launch has produced new areas for study but also has made clear topics where advocacy and action are needed, sometimes urgently. The establishment of a set of rights that can shape the development of the life sciences is one such urgent need.

Notably, there has been an increasing demand for scientific data and evidence for medical decision making and social and public policy creation—a trend that predates recent critiques of current federal research oversight in the United States. Those advocating for this primary role of scientific knowledge while correctly identifying the importance of the scientific verification process have not often highlighted the inherent difficulties, high costs, and slow progress of virtually all scientific enterprises. If we are going to depend solely on scientific data and evidence to change our lives and those of other needy world citizens, our cause is lost at its birth.

Rather, I would propose another menu to help garner benefit from the creativity and ingenuity that science brings to society. First, adopt the set of human rights offered in this book, recognizing their universality and effective appeal to the broad diversity of human interests. Second, educate scientists about these rights to help inform innovation. As a result, their work may more cautiously impinge on values we all hold dear. Collaboratively, scientists and other citizens could identify processes that need review and reinforcement, specify areas of policy inequities that require correction, and protect populations whose rights are now being violated. This is an environment in which the ingenuity and creativity of biotechnology will likely flourish. It is a context that could produce a newly responsible science that meets—rather than defines—human needs.

The cornerstone of this evolution of science is the focus on scientists as citizens and on the need for scientific knowledge in citizenship. More reflection, research, and education are surely needed. From this, or simply arising from demands for basic rights, new coalitions and types of political action may result—a framing of ingenuity and creativity by human rights and of knowledge in the service of human needs.

But this outcome is not certain. It will require thought, action, and goodwill. What I am convinced of is that dogma and outdated political stridency will not succeed, and self-reflection and modesty by scientists and their critics will likely be helpful. From this may emerge the strength to persuade or act against strong and entrenched forces.

Paul R. Billings
Council for Responsible Genetics
November 2004

NOTE

1. C. M. Christensen, R. Bohmer, and J. Kenagy, "Will Disruptive Innovations Cure Health Care?" *Harvard Business Review* (September–October 2000): 102–12.

Appendix: The Genetic Bill of Rights

$\mathcal{O}$ur life and health depend on an intricate web of relationships within the biological and social worlds. Protection of these relationships must inform all public policy.

Commercial, governmental, scientific and medical institutions promote manipulation of genes despite profound ignorance of how such changes may affect the web of life. Once they enter the environment, organisms with modified genes cannot be recalled and pose novel risks to humanity and the entire biosphere.

Manipulation of human genes creates new threats to the health of individuals and their offspring, and endangers human rights, privacy and dignity.

Genes, other constituents of life, and genetically modified organisms themselves are rapidly being patented and turned into objects of commerce. The commercialization of life is veiled behind promises to cure disease and feed the hungry.

People everywhere have the right to participate in evaluating the social and biological implications of the genetic revolution and in democratically guiding its applications.

To protect our human rights and integrity and the biological integrity of the earth, we, therefore, propose the Genetic Bill of Rights.

1. All people have the right to preservation of the earth's biological and genetic diversity.
2. All people have the right to a world in which living organisms cannot be patented, including human beings, animals, plants, microorganisms and all their parts.

3. All people have the right to a food supply that has not been genetically engineered.

4. All indigenous peoples have the right to manage their own biological resources, to preserve their traditional knowledge, and to protect these from expropriation and biopiracy by scientific, corporate or government interests.

5. All people have the right to protection from toxins, other contaminants, or actions that can harm their genetic makeup and that of their offspring.

6. All people have the right to protection against eugenic measures such as forced sterilization or mandatory screening aimed at aborting or manipulating selected embryos or fetuses.

7. All people have the right to genetic privacy including the right to prevent the taking or storing of bodily samples for genetic information without their voluntary informed consent.

8. All people have the right to be free from genetic discrimination.

9. All people have the right to DNA tests to defend themselves in criminal proceedings.

10. All people have the right to have been conceived, gestated, and born without genetic manipulation.

Issued by the Board of Directors of the Council for Responsible Genetics in 2000: Claire Nader (chair), Martha Herbert, Colin Gracey, Paul Billings, Philip Bereano, Debra Harry, Ruth Hubbard, Jonathan King, Sheldon Krimsky, Stuart Newman, Devon Peña, and Doreen Stabinsky.

Index

About the Editors

Sheldon Krimsky is professor of urban and environmental policy and planning and adjunct professor of public health and family medicine at Tufts University. He is a founding board member of the Council for Responsible Genetics. Professor Krimsky has published over 140 papers and reviews, and seven books including *Genetic Alchemy: The Social History of the Recombinant DNA*, *Biotechnics and Society: The Rise of Industrial Genetics*, *Agricultural Biotechnology and the Environment* with R. Wrubel, and *Science in the Private Interest*. A large segment of his published work is on the societal impacts and ethical implications of genetics and biotechnology.

Peter Shorett works as a freelance writer in Boston, Massachusetts. He writes in a variety of venues on the economics and politics of science and is a former director of programs at the Council for Responsible Genetics.

About the Contributors

Matthew Albright lives in Durango, Colorado, and is the author of *Profits Pending: How Life Patents Fail Science and Society*.

Joseph S. Alper is professor of chemistry at the University of Massachusetts at Boston and founding member of the Genetic Screening Study Group.

Philip Bereano is professor of technology and public policy at the University of Washington. He served as a civil society negotiator for the Cartagena Biosafety Protocol and is on the national board of the American Civil Liberties Union.

Richard Caplan is an environmental advocate for the U.S. Public Interest Research Group in Washington, D.C.

Marcy Darnovsky is associate executive director of the Center for Genetics and Society in Oakland, California.

Graham Dutfield is Herchel Smith Senior Research Fellow at the Queen Mary Intellectual Property Research Institute of the University of London.

Debra Harry is a Northern Paiute from Pyramid Lake, Nevada, and executive director of the Indigenous Peoples Council on Biocolonialism.

Martha R. Herbert is a pediatric neurologist and brain development researcher at the Massachusetts General Hospital and an assistant professor of neurology at Harvard Medical School.

Ruth Hubbard is professor emerita of biology at Harvard University and author of five books, including *Exploding the Gene Myth, Profitable Promises*, and *The Politics of Women's Biology*.

Jonathan King is professor of molecular biology at the Massachusetts Institute of Technology.

Jeroo Kotval is a consultant trained as a molecular geneticist with a long-standing involvement in ethical issues arising from the Human Genome Project.

Marc Lappé is executive director of the Center for Ethics and Toxics and author of several books, including *Chemical Deception, Evolutionary Medicine*, and *The Tao of Immunology*.

Paul Steven Miller is professor of law at the University of Washington School of Law and former commissioner of the U.S. Equal Employment Opportunity Commission.

José F. Morales is director of Public Interest Biotechnology and a postdoctoral associate in the Laboratory of Human Genetics and Hematology at Rockefeller University.

Peter J. Neufeld is co-founder and co-director of the Innocence Project at the Benjamin N. Cardozo School of Law in New York City.

Stuart A. Newman is professor of cell biology and anatomy at New York Medical College, where he directs a research program in vertebrate developmental biology.

Hope Shand is research director of the ETC Group, formerly known as the Rural Advancement Foundation International (RAFI).

Vandana Shiva is a world-renowned environmental leader, activist, and director of the Research Foundation for Science, Technology, and Ecology in New Delhi, India.

Doreen Stabinsky is professor of environmental politics at College of the Atlantic and a genetic engineering campaigner for Greenpeace International.

Sarah Tofte is program coordinator at the Innocence Project, where she directs the Innocence Network's national "Wrongful Convictions: Causes and Remedies" course for law and graduate students.

Brian Tokar is director of the Biotechnology Project at the Institute of Social Ecology in Vermont and author of four books, including the edited volumes *Redesigning Life?* and *Gene Traders*.

John Tuhey is counsel at Tellabs Operations, Inc., a global telecommunications firm.

Gregor Wolbring is a biochemist and adjunct assistant professor at the University of Calgary, and he is founder and coordinator of the International Network on Bioethics and Disability.